GCSE BITESIZE revision

Science
Foundation Level

Jane Vellacott
Ann Gregory
Mary Whitehouse

Published by BBC Worldwide Limited,
Woodlands, 80 Wood Lane, London W12 0TT

First published 2002
Reprinted July 2003, June 2004

© Ann Gregory, Jane Vellacott, Mary Whitehouse
/BBC Worldwide Limited, 2002.
All rights reserved.

ISBN: 0 563 501235

Colour reproduction by Tien Wah Press Pte Ltd, Singapore

Printed by Tien Wah Press Pte Ltd, Singapore

Acknowledgements
Material from the National Curriculum is Crown copyright and is reproduced by permission of the Controller of the HMSO.
Page 155 Paul and Lindamarie Ambrose / Getty Images (Milky Way); page 167, Gianni Tortoli / Science Photo Library (Turin Shroud); page 180–183, exam questions based on OCR 1794/5 1998, 1794/5 1999, 1794/5 2000, material used is © OCR accepts no responsibility whatsoever for the accuracy or method of working in the answers given.

Contents

Physics

Introduction

About Bitesize

GCSE Bitesize is a revision service designed to help you achieve success at GCSE. There are books, television programmes and a website, each of which provides a separate resource designed to help you get the best results.

TV programmes are available on video through your school, or you can find out transmission times by calling 08700 100 222.

The website can be found at:
http://www.bbc.co.uk/schools/gcsebitesize/

About this book

This book is your all-in-one revision companion for GCSE. It gives you the three things you need for successful revision:

1 Every topic clearly organised and clearly explained.
2 **The most important facts and ideas highlighted for quick checking** – in each topic and in the extra sections at the end of this book.
3 **All the practice you need** – in the check questions in the margins, the practice sections at the end of each topic, and the exam questions at the end of this book.

Each topic is organised in the same way:

- **The bare bones** – a summary of the main points, an introduction to the topic, and a good way to check what you know.
- **Key facts** highlighted throughout.
- **Check questions** in the margin – have you understood this bit?
- **Remember tips** in the margin – extra advice on this section of the topic.
- **Exam tips** in red – specific things to bear in mind for the exam.
- **Practice questions** at the end of each topic – a range of questions to check your understanding.

The extra sections at the back of this book will help you check your progress and be confident that you know your stuff.

Exam questions and model answers

- A selection of exam questions with the model answers explained to help you get full marks. There are also some questions for you to try, complete with answers at the end of the section.

Topic checker

- Quick questions in all topic areas.
- As you revise a set of topics, see if you can answer these questions – put ticks or crosses next to them.
- The next time you revise those topics, try the questions again.
- Do this until you've got a column of ticks.

Complete the facts

- Another resource for you to use as you revise: fill in the gaps to complete the facts. Answers are at the end of the section.

Last-minute learner

- The most important facts in six pages.

How to use this book

This book is divided into the three science subjects, which are sub-divided into units that cover the key GCSE topics of Science. If you have any doubts about which topics you need to cover, ask your teacher.

For many of the units, there are corresponding sections on the video. In such cases, it's a good idea to watch the video sequence(s) *after* reading the relevant pages, but *before* you try to work through or answer the practice questions. This is because the video sequences give you extra information and tips on how to answer exam questions. It's also a good idea to write the time-codes from the video on the relevant page(s) of the book – this will help you find the video sequences quickly, as you go over units again.

The most important and popular sections of the GCSE Science specification (regardless of exam board) are covered by the book – but BITESIZE Science doesn't aim to give total coverage of all topics. So it's important to carry on using your school textbook and your own notes. Because all the main types of GCSE Science questions you will be tested on in the exam are covered, the general tips and suggestions will be useful, even if some of your specific topics do not appear in the BITESIZE Science book. Remember, the skills are transferable to the content of any topic.

Taken together, this book, the BITESIZE *Check and Test* book, and the video cover all the main skills and contain all the core knowledge required in GCSE Science.

GCSE Science

The National Curriculum sets a Programme of Study with four Attainment Targets for Science.

Sc1 Scientific Enquiry is in two parts. **Ideas and evidence in science** considers the power and limitations of science and how different groups have different views about the role of science. There will be questions about these ideas in your written exams. **Investigative skills** are assessed by coursework, which you will have to hand in to your teacher for marking. This is a chance to gain as many marks as possible even before you go into the exam. This book concentrates on ways in which you can improve your marks in the exams.

Sc2 Life Processes and Living Things
You might also know this attainment target as Biology.

Sc3 Materials and their Properties
You might know this as Chemistry.

Sc4 Physical Processes This is also called Physics.

Each of these attainment targets is worth 25% of the marks towards your GCSE in Science.

GCSE exams

All the GCSE Science specifications cover the four Attainment Targets. There is some variety in the details of the specification. You should find out from your teacher which you are taking and obtain a copy.

Linear or modular?

Some GCSE science specifications are **linear**. This means that the exams at the end of the course, in June, will test your knowledge and understanding of all the science you have learned in your GCSE course. There may be three papers: one for each of Biology, Chemistry and Physics; or the papers may be integrated, asking questions about all three subjects on each paper. You need to make sure you know what your papers will be like.

If you are taking a **modular** specification you will have been doing short module tests during the course. The marks from these will contribute to your final GCSE grade. You will take one or two final exams at the end of the course. This book should help you revise for the module tests and final exams.

Single or double award?

Most students take Double Award Science. This means when the results come out you will get a double grade for science – CC or AA, for example. You need to know all the work in this book for Double Award Science. If you are taking Single Award, you should check with your teacher which topics you need to know.

Foundation Tier or Higher Tier?

There are two tiers of examination in Science. This book covers the work for the Foundation Tier papers. Students taking the Double Award Foundation Tier exam can achieve grades GG–CC. Even if you do really well on the Foundation Tier papers you can't achieve a better grade than CC. However, students that do really badly on the Higher Tier papers and don't get enough marks for a DD will be ungraded. You should discuss with your teachers and parents which is the best tier to enter. And when you get into the exam make sure you are given the right paper!

Picking up marks in exams

Follow these tips to make sure you get all the marks you deserve. The examiners cannot read your mind – they can only give you marks for what you actually put down on paper.

Read all the questions carefully, because they contain the clues to the answers.

Make sure you answer the question.

Diagrams – use a pencil to add to a diagram, following the instructions in the question.

Graphs
- Label the axes and show the scale and units you are using.
- Plot each point neatly with a cross.
- Draw a line or curve smoothly with a single line.

Chemical equations
- Write out the word equation first.
- Write down the formula for each substance in the word equation.
- Balance the equation.

Calculations
- You must show working for full marks.
- Write down a word equation to show the ideas you are using.
- Substitute the numbers you know into the equation.
- Work out the answer and show the units.
- Even if you can't do the arithmetic, you may get marks for writing down the equation and units of the answer.

The exam papers

Questions often start with information or diagrams, which will help you understand. Underline important words – they will help you in your answers.

The marks for each part of the question are given. The size of the space given for the answer and the number of marks give you a clue about how long your answer might be. Some questions ask you to write several sentences to explain your ideas. Marks will be awarded for using good English and the correct technical terms.

Planning your revision

Do you know when your exam is? How long have you got to revise? It is no good leaving revision until the night before the exam. The best way to revise is to break the subject up into BITESIZE chunks. That is why we have broken the GCSE science course into topics. There are 81 science topics to revise in this book. You need to work out how many topics you need to do each day to get it all done before the exam. Use the contents page to plan your revision. You could write the date by each topic, to show when you will revise it.

Of course, you have other subjects to revise too. It is often better to cover more than one subject in an evening – a change is as good as a rest! So how about planning all your revision by working out how much time you have before the exams start and then sharing the days out amongst your subjects? Don't forget to leave some time to relax too!

Revision tips

It's no good just reading this book; to learn the material and understand it you need to be active. Here are some ideas to try:

- At the end of each double page, close the book and write down the key facts from those pages.

- When there is a labelled diagram to learn, draw a copy of the diagram without the labels. Look at the labels in the book, close the book, label the diagram and then check how many you got right.

- 'Look, cover, write, check' is a good way of learning all sorts of things – including spellings, equations and formulae.

- Use the glossary to make some flash cards. Write a definition on one side of the card and the word on the other.

Look at the word – can you write a definition? Look at the definition. Which word is it? How do you spell it?

- Use flashcards to learn the equations. Include the units for all the quantities in equations.

- Use flashcards to learn the word equations for chemical processes. Write the word equation on one side and the balanced equation with symbols on the other.

- Revise with a friend – revision is more fun with a partner.

- Make a set of flash cards to fit in your pocket – great for the day when the bus gets stuck in a traffic jam or you have to wait for a dentist's appointment!

Life processes

THE BARE BONES
- ➤ Life processes are what distinguish living from non-living things.
- ➤ Living things show complex organisation at all levels, from molecules and cells right up to the structure of whole organisms.
- ➤ A living thing can be studied at different levels of complexity and organisation — from a single cell up to a whole organism.

A Life processes and body systems

1 Living things differ from non-living things because of the <u>life processes</u> that go on inside them.

2 By responding to changes, the body systems of larger organisms keep conditions inside them remarkably constant, and at the optimum for life processes to occur (*see also page 13*).

Remember
Plants make their own carbohydrate food through photosynthesis. Even so, plants also require minerals for nutrition, which they get from soil.

Life process	What happens in the body
respiration	energy transfers from food to the high-energy ATP molecule in cells
feeding	provides nutrition, including energy and raw materials
sensitivity	changes inside and outside the body are detected and the body can respond
movement	animals move from place to place to find the best conditions for life; plants move slowly by growing in a certain direction
reproduction	offspring are produced, to replace dying individuals in the population
growth	a permanent increase in size, due to an increase in the number and size of cells
excretion	getting rid of waste materials such as urea (in urine), carbon dioxide (a waste product from respiration) or oxygen (a waste from photosynthesis)

Q How might an organism's sensitivity help it to survive?

B   Cells, tissues and organs

1   Living things are made of <u>cells</u> (see page 10), which are the smallest units of living matter.

- Microscopic organisms may be made of one cell, but larger plants and animals are made of many.

- Different types of cells are specialised to carry out different functions within a living thing.

2   A group of cells of the same sort makes up a <u>tissue</u>. Together, different tissues make up an <u>organ</u>. An organ has a special structure to perform a particular function (see pages 10 and 11).

Cells

In order to grow, all living things have some **cells**, which can divide, increase in size, and then divide again. In this way, a single **fertilised egg cell** may become millions of cells.

Tissues

Tissues are collections of similar cells, which work as a team to do a particular job. For example, the outer layer of a leaf – the **epidermis**.

Organs

Organs are collections of tissues, which work as a team to do the same job. For example, the stem, root and leaves are all organs.

Q Suggest why larger plants and animals can carry out a greater variety of activities than one-celled organisms.

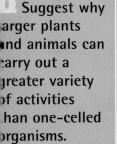

PRACTICE

1 What causes growth?

2 Which life process provides raw materials for building new cells?

3 Why must all living things carry out respiration?

4 How are plant and animal nutrition different?

5 Place these items in order of increasing size:
nerve tissue, light sensitive cell, the eye

Cells and cell activities

THE BARE BONES

➤ All cells have some structures in common, but there are important differences between plant and animal cells.

➤ The structure of a cell is related to its function.

➤ Cells divide during growth and reproduction.

A In plant and animal cells

A surface cell membrane surrounds the cytoplasm and nucleus.

KEY FACT

1 The membrane controls what moves in and out of cells.

• Only some particles can move through the membrane, depending on their size.

• Particles may diffuse through the membrane, following a concentration gradient, from a more concentrated area to a less concentrated area.

KEY FACTS

2 The cytoplasm is where chemical processes happen.

3 The nucleus is made of genetic material, such as DNA.

• DNA acts as a set of instructions for the cells.

• It is passed on to the next generation by inheritance.

• It gives a living thing its characteristics.

Plant cells also have a rigid cell wall, chloroplasts in the cytoplasm and an inner membrane surrounding a vacuole.

KEY FACTS

4 The cell wall is made of cellulose and supports the cell.

5 The chloroplasts are green because they contain chlorophyll, and they carry out photosynthesis.

6 The inner membrane surrounds a vacuole containing watery sap which helps plants to keep their shape.

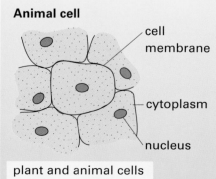

Animal cell
cell membrane
cytoplasm
nucleus

plant and animal cells have these features

Plant cell
cell membrane
cell wall
chloroplast
cytoplasm
inner membrane
nucleus

plant cells have these extra features

Q Can you describe what the nucleus, chloroplast and surface cell membrane do?

B Matching cell structure with function

Cells may look very different even though they share common features. This is because the structure of each cell suits the job it carries out.

cell	function and structure	
nerve	carries messages around the body; long, thin shape	
red blood	absorbs and caries oxygen; has a large surface area of cell membrane	
sperm	fertilises the egg cell; has a long tail allowing it to move	
root hair	absorbs water; has a large surface area of cell membrane	
egg	contains a lot of cytoplasm; if fertilised it develops into an embryo	
leaf	carries out photosynthesis; contains many chloroplasts	

Example question

Explain how the structure of an egg cell is related to its function.

1 Describe the structural feature.

2 Say how it helps the function.

The egg cell has a lot of cytoplasm because it contains food for a developing embryo.

Q How is the structure of a leaf cell suited to the job it performs?

C Cell division

Cell division that happens during growth or when replacing worn out cells is called <u>mitosis</u>.

Cell division that happens when sex cells are produced, is called <u>meiosis</u>.

For more on cell division and inheritance see page 50

Q Why do cells divide during growth?

PRACTICE

1 Name a feature of plant and animal cells that determines what the cells are like.

2 Which two features of plant cells help to provide support?

3 Suggest one reason why both a red blood cell and a root hair cell have a large surface area.

4 Which special feature of a plant cell allows it to photosynthesise its own food?

Body systems

THE BARE BONES

➤ The main body systems of a plant are the: • *shoot system* • *root system.*

➤ The main human body systems are the: • *breathing system* • *digestive system* • *reproductive system* • *blood system* • *skeletal system* • *urinary system* • *nervous and chemical coordination systems.*

A The plant body

KEY FACT

1 The shoot system is made up of the leaves, stems and flowers.

• The main function of leaves is to carry out photosynthesis. Leaves convert simple raw materials from the air and the soil into carbohydrate food, which can be stored or built into other substances.

• Stems transport materials between the shoot and root system, as well as supporting the shoot system above ground level.

• The flowers are the reproductive structures, where seeds form.

KEY FACT

2 The root system absorbs water and minerals, and anchors the plant.

Remember
Because plants can make their own food by photosynthesis, they are known as producers. Producers are always at the start of a food chain (see page 64/65).

Q Which structure in plants is most concerned with nutrition?

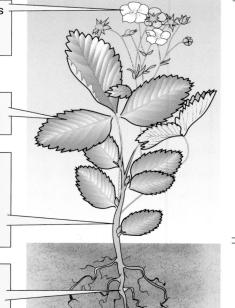

Flowers contain reproductive structures which produce sex cells.
After pollination, seeds form here.

Leaves are photosynthetic organs.
They make food.

Stems contain transport cells, carrying:
• food from the leaves to the buds and roots
• water and minerals from the roots to the shoot system

The **root** system:
• absorbs water and minerals
• anchors the plant
• may have root nodules where nitrogen compounds form

shoot system

root system

B   Human body systems

EY FACT

> The <u>body systems interact</u> to bring about the <u>healthy functioning</u> of the <u>whole body</u>.

Make sure you know the difference between breathing and respiration. Breathing is about getting oxygen from air into the body, and respiration is the transfer of energy from food to cells.

Example question

Which body systems are involved in supplying energy to cells?

the breathing system – gets oxygen into the body

the digestive system – gets food (glucose) into the body

the blood system – delivers oxygen and glucose to cells

life process respiration – transfers energy from food to cells.

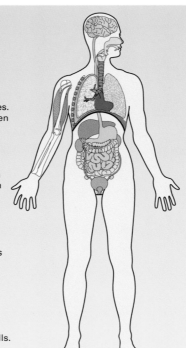

The **nervous system** is composed of sense or *receptor cells* which detect changes inside and outside the body. This information is *processed* by the brain, which causes muscles and glands to *respond* to change.

The **breathing system** involves the lungs, diaphragm, ribs and rib muscles. Air moves in and out of the lungs when we breathe because of *air pressure changes* inside the chest cavity.

The liver is important, as it:
• makes bile which helps fat digestion
• controls the level of digested food in blood.

The **blood system** is the major transport system. It is made up of:
• the heart (a pump)
• tubes (blood vessels such as arteries and veins)
• blood (which carries materials)
Blood is also important in fighting disease and healing wounds

The **reproductive system** contains the sex organs which produce sex cells.

The **skeletal system** is made of bone, cartilage, tendon and muscle. It:
• gives protection to delicate organs
• supports soft tissues
• allows us to move

The **digestive system** is basically a tube from mouth to anus, with different features along its length. Food is broken down (*digested*) into smaller particles and *absorbed* into the blood. Undigested food passes out of the body as faeces.

The **urinary system** consists mainly of the kidneys which:
• excrete urea in urine
• balance the amount of water and salts. Urine is stored in the bladder before passing out of the body.

The **chemical control system** involves *glands* which make *hormones*. Hormones are carried by blood and act on particular *target organs*. This helps the body respond to changes in body conditions.

Why is a blood system needed in a multi-celled organism?

PRACTICE

1 What is the advantage of the stem holding the shoot system well above ground?

2 In which direction does water pass through a plant?

3 Which part of a plant contains the male and female sex cells?

4 Which human body system has a major role in transport?

5 Immunity gives protection against infection. Which body system helps give us immunity?

6 Which body systems interact to allow the body to respond to change in a coordinated way?

Human nutrition

➤ We get nutrition or nutrients from our food. This provides:
 • *raw materials to build new cells* • *energy for life processes*
➤ People need a diet that suits their lifestyle, which can change during their lifetime.

A   The main food types

KEY FACT

1   A balanced human diet includes carbohydrates, proteins, fats, water, minerals and vitamins. Variety is important.

Remember
Fibre, starch and sugar are carbohydrates [listed here in order of decreasing size]. Only sugar is soluble, which is why the body digests larger particles.

Carbohydrates are important as they are needed to release energy around the body, in order for all the different processes to take place.

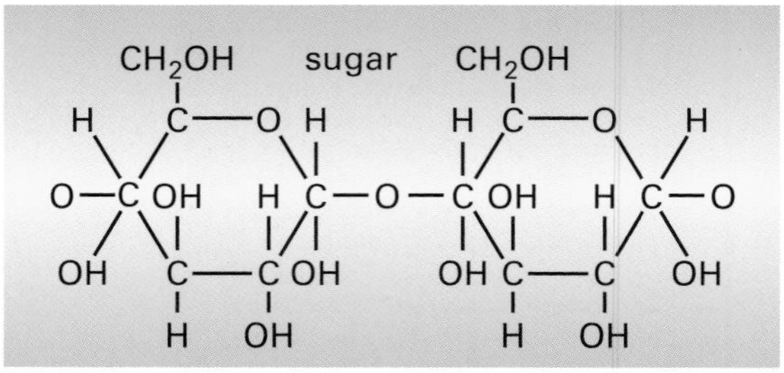

Amino acids are the building blocks of proteins, such as hormones, enzymes and complex molecules, such as DNA.

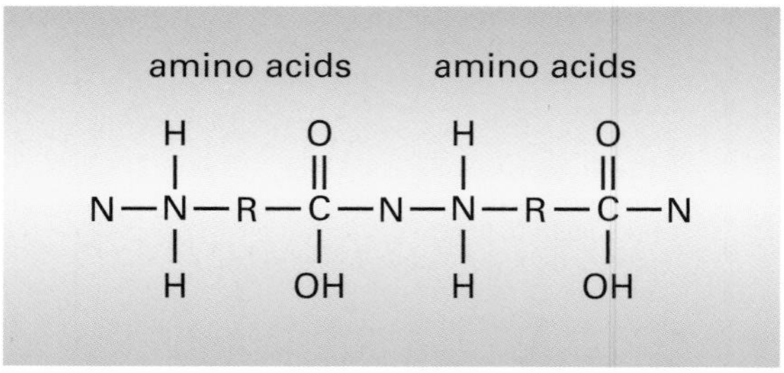

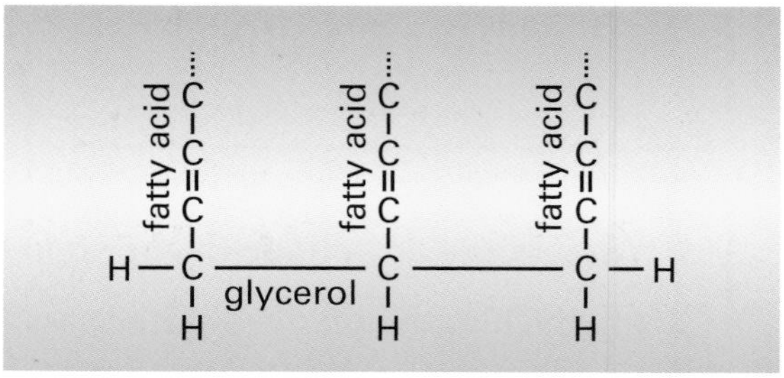

Q Can you name a simple carbohydrate which passes directly into the bloodstream?

Three fatty acids and glycerol have combined in this structure to form a triglyceride. Triglycerides are important in maintaining body temperature and as energy storage.

KEY FACT

2   Minerals and vitamins keep us healthy, and some are needed for enzymes to work. Some vitamins are destroyed by cooking, which is why we need to eat raw vegetables or fruit.

A

how the body uses minerals and vitamins	food containing the mineral or vitamin
iron – used to make new red blood cells	meat, spinach
calcium – needed to build strong teeth and bones	milk, eggs
vitamin C – healthy skin and blood vessels	tomatoes, cherries, oranges, lemons
B vitamins – for the nervous system to work properly	yeast, liver, wholemeal bread

 Suggest a reason why some people take vitamin supplements.

B   *Diet for a healthy lifestyle*

1   It is important to <u>match diet with lifestyle</u>, according to:
• <u>age</u> and <u>gender</u> • how <u>active</u> someone is
• some <u>medical conditions</u>.

KEY FACT

Remember energy is measured in joules. 1 kilojoule (kJ) = 1000 joules.

If men and women are equally active, why do men need to eat more than women?

child aged 6 7500kJ	someone doing physical work needs more energy than someone sitting in an office	high cholesterol: eat low fat foods
girl aged 12–15 9600kJ		diabetic: control sugar intake carefully see page 37
woman 9500kJ	adults need more energy than children and men need more energy than women due to larger body mass; elderly people are less active than younger people	high blood pressure: avoid salty foods see page 27
man 11 500kJ		kidney not working well: eat less protein see page 42

PRACTICE

1 Which two food types provide most energy in our diet?

2 Why is it important to have some fat in our diet?

3 Why are salads and lightly-cooked vegetables good for you?

4 If someone is actively body building, which food type might they need more of in their diet?

5 Name a protein source suitable for a vegetarian.

6 Why do elderly people need smaller meals than young adults?

7 What are the three most likely reasons for someone being overweight?

Human digestion

THE BARE BONES

➤ Food is broken down into smaller particles in the alimentary canal.

➤ Enzymes are mainly responsible for digestion.

➤ Digested food particles are absorbed into the bloodstream and carried to all the cells of the body.

A The structure of the gut

Q Which organ other than the alimentary canal is important in digestion?

Remember
The small intestine comes before the large intestine, and is responsible for digestion and absorption of digested food using the villi.

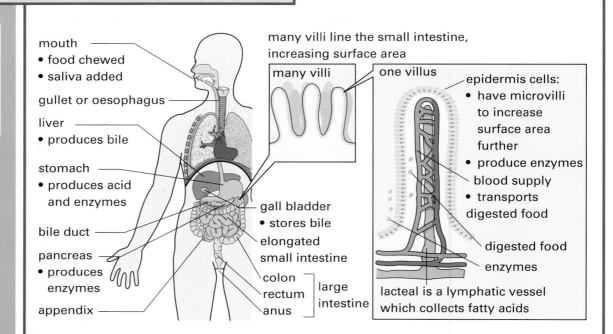

mouth
• food chewed
• saliva added

gullet or oesophagus

liver
• produces bile

stomach
• produces acid and enzymes

bile duct

pancreas
• produces enzymes

appendix

gall bladder
• stores bile

elongated small intestine

colon
rectum } large intestine
anus

many villi line the small intestine, increasing surface area

many villi

one villus

epidermis cells:
• have microvilli to increase surface area further
• produce enzymes

blood supply
• transports digested food

digested food

enzymes

lacteal is a lymphatic vessel which collects fatty acids

B Food processing

Q What is the name for the passage of food through the gut?

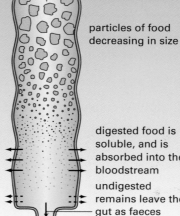

■ teeth chop and grind food (called physical digestion)

■ saliva softens food and adds enzymes

■ enzymes speed the breakdown of food particles

PERISTALSIS
food moves through the gut because muscles in the wall contract to squeeze it along

■ absorption of digested food

■ water reabsorbed

food enters gut

particles of food decreasing in size

digested food is soluble, and is absorbed into the bloodstream

undigested remains leave the gut as faeces

mouth (slightly alkaline) produces:
• saliva containing carbohydrase called amlyase, which begins digestion of starch to sugar

stomach (acid) produces:
• protease digests proteins to amino acids
• acid kills bacteria and helps to break down proteins to amino acids

small intestine (alkaline) and the pancreas produce:
• carbohydrate
• protease
• lipase digests fat to fatty acids and glycerol

liver adds bile to the small intestine:
• breaks up larger fat droplets
• neutralises acid

c About enzymes

1 Enzymes speed up chemical reactions, and are called <u>biological catalysts</u>.

2 Enzymes are protein molecules, each has a <u>unique shape</u>.

- The **shape** of an enzyme is **very important to how it works**.
- Molecules **temporarily bind** to a place on the enzyme called the **active site**.
- Molecules react at the active site, either by breaking into smaller molecules or by joining into larger ones.

3 If the shape of an enzyme changes, it does not join with other molecules and cannot speed up reactions.

These conditions affect the shape of enzymes:

- **high temperature** (>50°C), because the enzyme's protein structure breaks down
- **change in pH**, because pH affects bonds in the protein molecule.

Q Why does mylase in saliva top working when it reaches he stomach?

You will need to know the general names for digestive enzymes. The names of the enzymes are very like the food they digest – just add 'ase' to the first part of the food name:
- **carbohydrate** is digested by **carbohydrase**
- **protein** is digested by **protease**
- **lipid (fat)** is digested by **lipase**.
  You need to know one example of a carbohydrase, which is **amylase**.

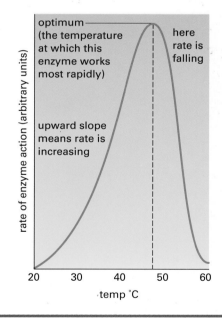

rate of enzyme action (arbitrary units)

optimum (the temperature at which this enzyme works most rapidly)

here rate is falling

upward slope means rate is increasing

temp °C

20   30   40   50   60

1 Where in the gut is most protein digested?

2 Name three places where the carbohydrase enzyme called amylase is produced.

3 Describe three ways in which the gut is adapted for maximum surface area.

4 What happens to digested food?

5 Why do enzymes stop working above 50°C?

The heart

THE BARE BONES

➤ The heart is mainly made of muscle. It beats throughout life, pumping blood around the body.

➤ Arteries, veins and capillaries are the blood vessels which form the circulation system to all cells.

A The heart

The heart is made up of four chambers: <u>two atria</u> and <u>two ventricles</u>.

Remember
Atrium = singular, atria = plural. Another word for atrium is auricle.

1 One atrium and one ventricle make up each side of the heart.

2 The left hand side of the heart (LHS) works separately but simultaneously to the right hand side (RHS).

3 The LHS of the heart pumps blood to most of the organs and tissues of the body, and the RHS pumps blood to the lungs.

4 The largest artery in the body is the aorta, which carries blood from the LHS of the heart to most of the tissues and organs of the body.

5 Valves control the direction of blood flow through the heart.

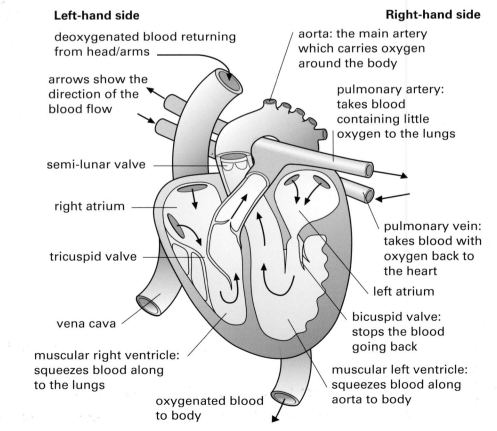

Left-hand side

deoxygenated blood returning from head/arms

arrows show the direction of the blood flow

semi-lunar valve

right atrium

tricuspid valve

vena cava

muscular right ventricle: squeezes blood along to the lungs

oxygenated blood to body

Right-hand side

aorta: the main artery which carries oxygen around the body

pulmonary artery: takes blood containing little oxygen to the lungs

pulmonary vein: takes blood with oxygen back to the heart

left atrium

bicuspid valve: stops the blood going back

muscular left ventricle: squeezes blood along aorta to body

Q Can you name the four chambers of the heart and the largest artery and vein?

B   The heart beat

A heart beat happens when the <u>heart muscle contracts</u>, squeezing blood through the heart and out into the arteries. In between each beat the heart muscle relaxes.

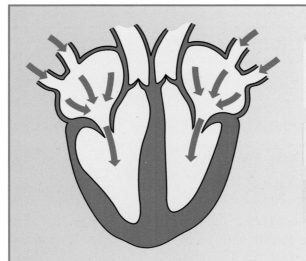

- the atria are relaxed and fill with blood, which enters through the large veins

- then the atria contract and blood passes into the ventricles

- valves at the base of the large arteries are closed

Make sure you understand the difference between atria and ventricles, arteries and veins.

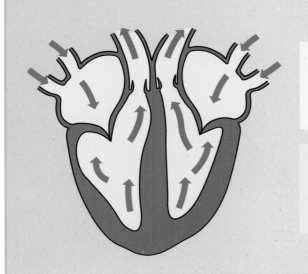

- as the ventricles fill, blood presses against the valves between the atria and the ventricles, closing them

- the ventricles contract strongly, pumping blood into the arteries and pushing open the valves at the base

Q Suggest a reason why the heart beat is automatic.

1 The muscle on the LHS of the heart is slightly thicker than on the right. Suggest a reason why.

2 How does the heart muscle bring about a heart beat?

3 Where are the valves in the heart? Why are they necessary and when do they close?

Circulation and the blood system

THE BARE BONES

➤ Blood is a liquid containing dissolved substances, plasma proteins, platelets and cells.

➤ Arteries, veins and capillaries are different types of blood vessel.

A Circulation and blood vessels

KEY FACT

Arteries carry blood to organs from the heart, and veins carry blood from organs back to the heart. Capillaries are tiny vessels which come into close contact with all cells.

Remember
Tissue fluid is the liquid in-between the cells that make up tissue and organs.

The structure of blood vessels

feature	structure	O$_2$ blood level	blood pressure	speed and direction
artery	thick, muscular, elastic wall with narrow space inside; resists high pressure and pulls back into shape	high	high	high; blood flows towards organs and away from the heart
capillary	thin wall only one cell thick, tiny and passing close to cells; wall leaks	higher near an artery and lower near a vein		very slow, so more time for materials to exchange between blood and tissue fluid
vein	thinner wall than artery with a larger space inside; valves stop back flow	lower than an artery	lower than an artery	slower than an artery; blood flows from towards heart

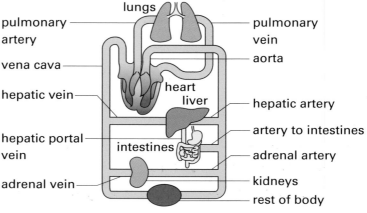

Q Why is the circulatory system in humans called a 'double circulation'?

B Components of blood

1 The main components of blood are:
* the liquid part of blood called <u>plasma</u>
* red and white blood <u>cells and platelets</u>.

2 Haemoglobin is found in red blood cells.

* In the lungs where there is a more oxygen, haemoglobin combines with oxygen to form oxyhaemoglobin.

* In the tissues where oxygen is being used up, oxyhaemoglobin splits up to release oxygen to the cells.

red blood cells
flattened, dented discs, which:
* carry oxygen to the cells

platelets are cell fragments, which are important in blood clotting

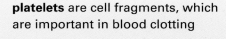

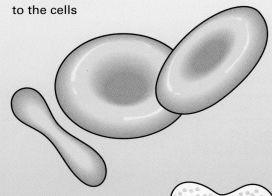

there are several types of
white blood cells:
* some engulf and destroy bacteria

* some produce antibodies and anti-toxins
* neutralise toxins

plasma is the liquid part of the blood, which carries:
* carbon dioxide to the lungs
* digested food
* waste materials such as urea

Which parts of blood are most concerned with transport?

1 Why are there pressure and oxygen level differences in the blood that flows in the aorta and the vena cava?

2 What makes haemoglobin pick up oxygen, and oxyhaemoglobin split up?

Health and disease

THE BARE BONES

➤ The body's main defences against infection include the skin and the blood system.
➤ Poor health may be caused by microbes.
➤ There are some diseases that, as yet, cannot be cured.

A Poor health

KEY FACT

> The causes of poor health include: infection, injury, inherited conditions, lifestyle and mental illness.

There are many different reasons for ill health, such as:

1 infections – a cold is caused by a virus, a sore throat is caused by a virus or bacterium and athletes' foot is caused by a fungus

2 injury – the liver might be damaged in an accident

3 an inherited condition – cystic fibrosis

4 a lifestyle habit such as drug abuse

5 mental illness – paranoia.

Q Suggest how poor health affects the quality of life.

B Microbes and disease

KEY FACT

> **1** Not all micro-organisms cause disease, but some do.

Bacteria	Fungi	Viruses
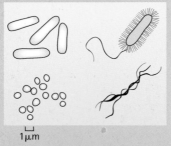	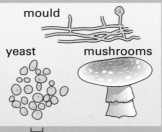	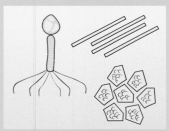
1μm	5μm	$1\mu m = \frac{1}{1000}\,mm$

Bacteria
- single–celled
- variety of shapes
- microscopic: a light microscope is needed to see them
- genetic material, but no real nucleus
- may be disease–causing (pathogenic)
- feed in a variety of ways

Fungi
- mainly many–celled, but yeast are single–celled; moulds are thread–like and form spores; some form mushrooms
- bring about decay by decomposing dead remains of living things
- may be pathegenic
- microscopic, but many may be seen without a microscope

Viruses
- very small, too small to be seen with a light microscope – need to use an electron microscope
- viruses have genetic material, but no real nucleus
- many are pathogenic
- can only live and reproduce inside other cells

Q Can you list the three types or micro-organisms shown here in order of size, smallest first?

B

- As yet there is no 'cure' for HIV because the virus particles are inside white blood cells, where they are protected from our immune system.

Y FACT

2 HIV is a virus that enters human cells, including white blood cells that normally protect us against infection.

- The virus particles take over white blood cells, using them to make more HIV particles, and destroying the white blood cells.

C Body defences against disease

Y FACTS

1 The main defence against disease is the <u>skin</u>, which is a <u>barrier to infection</u>. Skin <u>replaces itself continuously</u> allowing a wound to heal.

2 The breathing tubes (trachea, bronchi and bronchioles) are lined with cells which have <u>cilia</u> and produce <u>mucus</u>. Dirt and micro-organisms are swept away from the lungs in mucus, by the cilia.

3 Blood clotting stops blood loss, and also stops germs from entering the body. A cut blood vessel gets plugged by a clot, which soon turns into a scab. This is the start of wound healing.

What type of health condition is each of the following:

an earache?
a broken leg?
an addiction to tranquillisers?
seriously depressed?
haemophilia?

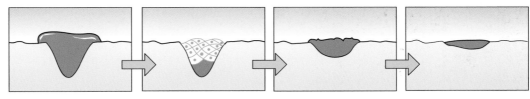

| Blood emerges from wound and fills the space. | The blood begins to clot on contact with the air. | A protective scab forms over the top of the wound. | Beneath the scab, the wound begins to heal. |

PRACTICE

1 Measles is caused by a virus. If you have measles, what type of disease is it?

2 How can our natural defences against measles be improved by medical means?

3 Describe three ways in which white blood cells protect against disease.

Immunity

THE BARE BONES
➤ Immunisation is a way of protecting ourselves against disease.
➤ Lifestyle has a significant effect on health.

A *The body's natural immunity*

1 Some white cells recognise <u>antigens</u> and produce <u>antibodies</u> to match. This is called an <u>immune response</u>.

- An antigen is a protein or 'non-self' cell, such as a bacterium, that enters the body.
- Antibodies lock onto antigens, damaging them or letting white cells engulf them.
- Some white blood cells produce anti–toxins to neutralise the bacterial toxins.

Remember
Immunisation, or vaccination, stimulates the body's immune response to protect against disease.

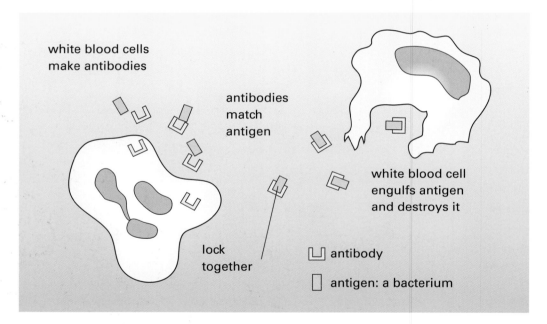

white blood cells make antibodies

antibodies match antigen

white blood cell engulfs antigen and destroys it

lock together

⊔ antibody

▯ antigen: a bacterium

Q Haemophilia is an inherited disease in which the blood lacks clotting factors. Suggest why this can be life-threatening.

2 Saliva and tears contain anti-bacterial substances to help prevent infection.

B   Immunisation

EY FACT

We can use antigens which are not dangerous to health to cause an immune response. This can give protection for the future.

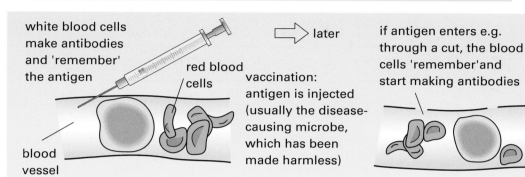

white blood cells make antibodies and 'remember' the antigen

red blood cells

blood vessel

later

vaccination: antigen is injected (usually the disease-causing microbe, which has been made harmless)

if antigen enters e.g. through a cut, the blood cells 'remember' and start making antibodies

Case study

How can we stay healthy?

There is no doubt that we influence our health by the way we live. What do you think about what these people say about staying healthy?

'My little dog lets me know when it's time for our daily walks, so I keep a lot fitter than many people my age'.

'If I don't eat fibre with my breakfast I get constipated!'

'Breathing in smoke lets harmful chemicals get into blood, and cells in the lungs get damaged.'

'A balanced diet is really important. Let's face it, if you only eat chips and sweets all the time, you'll probably get fat'.

'I'm mad on fruit and veg big time – if I eat plenty I get less spots and feel more energetic.'

'All drugs such as medicines, caffeine and alcohol, or illegal substances such as heroin, change the way the body works. Drugs can have harmful effects and can be dangerous to health. Of course medicines are controlled, so they are safe to use as instructed'.

"At the office I'm sitting most of the day, so getting to the gym after work is great. It keeps me supple and strengthens my muscles'.

Q Which blood cells are involved in immune response?

PRACTICE

1 Describe three main factors which are important for a healthy lifestyle.

2 Suggest one type of data which would prove a link between smoking and disease.

The breathing system

THE BARE BONES

➤ The lungs are the main organs of the breathing system.
➤ Breathing happens when muscles pull on the ribs and the diaphragm.
➤ Air moves in and out of the lungs by a difference in air pressure.
➤ Gas exchange happens in the air sacs (alveoli).

A The structure of the breathing system

KEY FACT

Remember
Gas particles diffuse along a gradient, from an area where there is more of them to an area where there are fewer.

> Living things need to exchange gases with their surroundings.

1 Plants exchange gases by diffusion, mainly through pores in the leaf surfaces.

2 Animals are so active that they need a greater supply of oxygen (from air) than they could get by diffusion alone.

3 The breathing system is adapted for getting air in and out of the body fast.

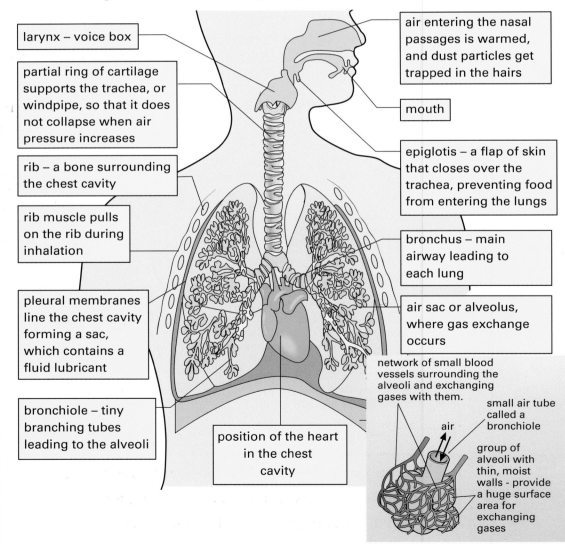

larynx – voice box

partial ring of cartilage supports the trachea, or windpipe, so that it does not collapse when air pressure increases

rib – a bone surrounding the chest cavity

rib muscle pulls on the rib during inhalation

pleural membranes line the chest cavity forming a sac, which contains a fluid lubricant

bronchiole – tiny branching tubes leading to the alveoli

air entering the nasal passages is warmed, and dust particles get trapped in the hairs

mouth

epiglotis – a flap of skin that closes over the trachea, preventing food from entering the lungs

bronchus – main airway leading to each lung

air sac or alveolus, where gas exchange occurs

position of the heart in the chest cavity

network of small blood vessels surrounding the alveoli and exchanging gases with them.

air

small air tube called a bronchiole

group of alveoli with thin, moist walls - provide a huge surface area for exchanging gases

Q Name the air tubes which lead from the nose and mouth to inside the lungs.

B Gaseous exchange and breathing movements

1 Gas exchange happens in the air sacs (alveoli).

- Oxygen **dissolves** in surface moisture and **diffuses** through the wall of the air sac into a blood capillary.
- Carbon dioxide diffuses from blood into the air sac, and is breathed out.
- Inhaled air contains: 78% nitrogen, 21% oxygen, 0.04% carbon dioxide, 1% noble gases.
- Exhaled air contains: 78% nitrogen, 18% oxygen, 3% carbon dioxide, 1% noble gases.

2 Breathing is an <u>active process</u> that uses energy because we need to move muscles to make it happen. Breathing is <u>controlled automatically</u>, by the nervous system.

3 The direction of air flow depends on the <u>difference in pressure</u> between the inside of the lungs and outside the body.

C Smoking and the breathing system

Smoking damages health and the breathing system. The main reasons are:

effect	problem it causes
smoke particles and chemicals damage the cilia on the cell surfaces	cilia cannot sweep out dirt that collects in mucus; fluid collects in lungs causing breathlessness
smoke contains tar which is carcinogenic	smokers are more likely to get lung cancer
nicotine in smoke • causes blood vessels to contract • is addictive	• raises blood pressure and damages arteries so blood clots may form, and a stroke is more likely • it is hard to give up smoking
carbon monoxide combines with haemoglobin	blood carries less oxygen

1 In what ways are the alveoli well-adapted for gas exchange?

2 How do breathing movements bring about air flow?

3 Why does smoking tend to raise blood pressure?

THE BARE BONES
➤ Respiration transfers energy from a food source to living cells.
➤ It occurs aerobically using oxygen, transferring lots of energy.
➤ It occurs anaerobically without oxygen, and transfers less energy.

A Aerobic respiration

KEY FACT

1 All <u>living cells</u> in the body carry out respiration.

2 During aerobic respiration, chemical reactions occur that:
* use oxygen and glucose
* break down glucose to produce carbon dioxide and water as waste products
* transfer energy to the cell.

3 glucose + oxygen → carbon dioxide + water (+ energy transferred)

Q Why do cells carry out respiration?

B Anaerobic respiration

KEY FACT

1 During <u>vigorous exercise</u>, muscle tissue uses oxygen very quickly – supply can not meet the demand. When there is <u>not enough oxygen</u>, <u>anaerobic respiration</u> occurs.

* Glucose is broken down into lactate, and some energy is transferred to the cell.
glucose → lactate (+ some energy transferred)

2 When vigorous exercise stops, more oxygen becomes available to the muscle cells.

* The cells use the oxygen to break down the lactate to carbon dioxide and water.
* If there is a build-up of lactate in muscle, the muscle cells have an oxygen debt.

3 Other organisms (e.g. yeast) carry out anaerobic respiration. This is called fermentation and forms alcohol as a waste product.

glucose → carbon dioxide + ethanol + some energy transferred

Q Why do people breathe faster when they exercise?

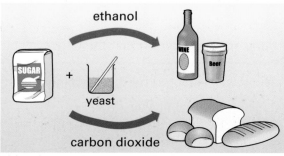

Brewing
• ethanol (alcohol) is important in brewing
• carbon dioxide gas escapes when making wine or beer

Baking
• gas bubbles make the bread dough rise
• the ethanol evaporates during baking

c Why cells need energy

building up particles and cells

moving

keeping warm

You will often be asked to find data and use it to answer questions.

How is energy used in cells? You should be able to think of three different reasons.

Example A shrew is a small mammal. The graph shows the link between body mass and the rate at which oxygen is used up for different species of shrew.

- A note on units: cm^3/g/h means 'cm^3 per gram of body mass per hour'.

- Make sure you can read graphs, for instance:

Q Which species of shrew uses $7cm^3$/gram/h of oxygen?

A The unit in the question is on the y-axis. Read across from the y-axis at 7 to get the answer, which is *Monterey shrew*.

- You need to be able to describe a trend, for instance:

Q Describe the relationship of body mass of shrews to the rate they consume oxygen.

A The bigger the shrew, the less oxygen it uses per hour.

- Make sure you know how to interpret data, by saying what it means, for instance:

Q Suggest a reason for the trend you described.

A Bigger shrews lose heat more slowly than smaller ones, so they do not need to carry out respiration as fast, and they use less oxygen.

PRACTICE

1 Which of the waste products of respiration is breathed out?

2 Describe three ways in which aerobic and anaerobic respiration are different.

3 In what conditions do muscle cells build up an oxygen debt?

4 Standing still uses around 7kJ/min. How much energy does a boy need to stand for half an hour?

The basics of sensitivity

THE BARE BONES

➤ Survival depends on being sensitive to changes inside and outside the body, and responding appropriately.

➤ The nervous system gives us sensitivity.

➤ The brain acts as a central processing unit for the nervous system.

A Neurones

KEY FACTS

1 A <u>stimulus</u> is a condition detected by the body, e.g. temperature of the surroundings, or sugar level in blood.

2 A <u>receptor</u> is a part of the body which <u>detects a stimulus</u>.

The receptor could be:

• a single nerve cell (a sensory **neurone**), such as a pain receptor in the skin, sound or position receptor in the ear, or a taste receptor in the tongue

• a sense organ (like the nose) with many smell receptor cells; the eye with many photoreceptors in the retina.

KEY FACT

3 Sensory and motor neurones have a different structure, which is related to their function.

• Sensory neurones are linked to receptors.

• Motor neurones are linked to effectors.

Sensory neurone

receptor detects a stimulus eg. touch

impulse travels to central nervous system

Motor neurone

cell body contains most of the cytoplasm

axon is a long fibre, connecting two areas of the body

impulse travels from central nervous system to an effector

nerve endings connect with an effector, such as a muscle

Q What is the difference in function between a receptor and an effector?

B Organisation of the nervous system

1 The brain and spinal cord make up the <u>central nervous system</u>. The brain is a processing centre for information detected by receptors. It sends messages to effectors.

2 An effector is the part of the body that responds when instructed by the brain e.g. a muscle which moves, or a gland which produces a hormone.

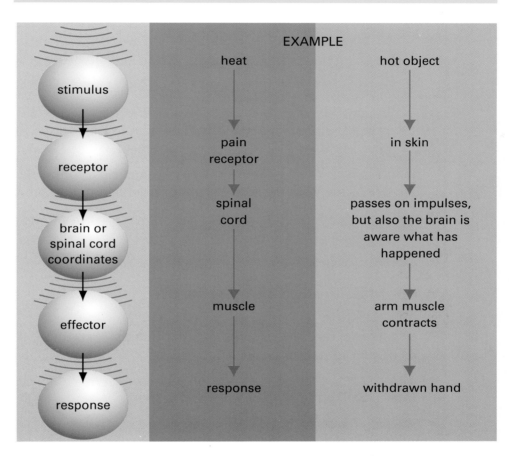

EXAMPLE

	heat	hot object
stimulus	↓	↓
receptor	pain receptor	in skin
brain or spinal cord coordinates	spinal cord	passes on impulses, but also the brain is aware what has happened
effector	muscle	arm muscle contracts
response	response	withdrawn hand

An example for you to try

Judy was tired after a day's shopping, and couldn't wait to sit down with a cup of tea. She sank into the sofa and kicked off her shoes. Just then she got a text message on her mobile and, as she stood up to get her phone, she experienced a stabbing pain in her foot. Then she remembered her daughter had been sewing on a button the day before.

Why is it useful for the brain to be aware of an action that has taken place?

PRACTICE

1 How is a sensory neurone adapted to its function?

2 What is the job of the fatty sheath around the axon of a neurone?

3 In what ways does the central nervous system act as a processing centre?

THE BARE BONES

➤ The body's responses to change may be voluntary or automatic.
➤ Reflex actions happen fast and automatically.
➤ Drugs change body function and may affect the nervous system.

A Responding and reacting

KEY FACT

1 When a <u>receptor</u> is stimulated it <u>sends messages</u> in the form of <u>electrical impulses</u> along a <u>sensory nerve</u> to the central nervous system.

- The response may be voluntary because the brain decides which response is best. This type of response is learned e.g. how to ride a bike or perform gymnastics.

- However, a reflex happens automatically. This type of response does not have to be learned. We blink when an object comes close to the eye unexpectedly, and the size of the pupil in the eye changes according to how much light enters it.

KEY FACT

2 In a simple reflex action, the nerve impulses pass from a receptor along a sensory nerve to the central nervous system, then along a motor neurone to a muscle or gland.

A new-born baby has reflex actions, which do not need to be learned:

- on touching its cheek, it turns its head towards its mother to search for the breast

- grab reflex – a baby throws its arms and tries to grab on to support when startled.

- stepping reflex – when its body mass is supported, but feet touch a firm surface.

- recoil reflex – when a wasp stings your hand you will automatically recoil your arm.

Q Why is it illegal to drive a car once the level of alcohol in the blood has risen above a certain level?

B Drugs and the nervous system

1 Drugs are found in food materials, tobacco, medicines and substances of abuse. Drugs • change how the body works • may affect how we behave • may cause harm.

Drug	Effects
solvents	affect behaviour; and may damage lungs, liver and brain
alcohol	affects behaviour, may lead to lack of judgement, unconsciousness; slows down reactions; may cause liver damage; can be addictive
nicotine	speeds up the heart rate; causes artery walls to contract, raising blood pressure; can be addictive

2 The abuse or misuse of any drug is potentially dangerous.

- Over-use of antibiotics has led to some bacteria becoming resistant to antibiotics.
- Caffeine is a stimulant, which speeds up heart rate and raises blood pressure.
- An overdose of painkillers can cause liver damage and be fatal.

Case study Quick reactions?

In an experiment to test reaction times, some university students were told to press a buzzer when they heard a bell ring. Their reaction times were measured. Then they drank some beer and were tested again. Here are the results showing that their reaction times were affected by drinking beer, and that some students recovered sooner than others.

Student	reaction time/milliseconds				
	before drinking beer	1h after drinking beer	2h after drinking beer	3h after drinking beer	4h after drinking beer
Manjit	80	125	95	79	80
Lisa	90	160	100	95	90
Darren	70	150	120	100	90
Jo	85	110	84	85	85

Alcohol acts s a sedative. ow does the ata support his fact?

PRACTICE

1 Which of these reactions are likely to be voluntary responses, and which are likely to be reflexes?
- moving your bare foot off a sharp stone
- playing a piano • 'crying' when you peel onions
- catching onto a rail to save yourself from falling • speaking

2 Which student in the case study above recovered the slowest? Suggest why.

The eye and sight

THE BARE BONES

➤ The eye is a specialised sense organ, which is designed for sight.
➤ The retina contains photoreceptors, which are sensitive to light.
➤ The eye can adjust to different light intensities.

A The eye

KEY FACT

1 The eye's structure allows <u>light reflected from objects</u> to enter it, which is detected by receptors in the retina.

Remember
The axons of sensory neurones collect to form the optic nerve, which passes directly to the brain. It is a short nerve, so we get visual information very quickly.

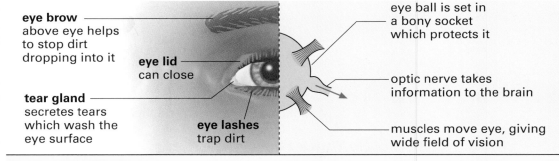

eye brow above eye helps to stop dirt dropping into it

eye lid can close

tear gland secretes tears which wash the eye surface

eye lashes trap dirt

eye ball is set in a bony socket which protects it

optic nerve takes information to the brain

muscles move eye, giving wide field of vision

tough outer coat, called the sclerotic layer or **sclera**
choroid layer, contains blood supply
retina, contains the light-sensitive receptor cells
ciliary muscle, which controls the shape of the lens
lens, made of a gel-like substance
cornea – transparent part of the outer coat, which helps in focusing
iris, controls amount of light entering eye
ligaments hold the lens

KEY FACT

2 The <u>cornea</u> and the <u>lens</u> focus images from near or distant objects.

An image is formed when light comes to focus on the retina. The image is interpreted by the brain.

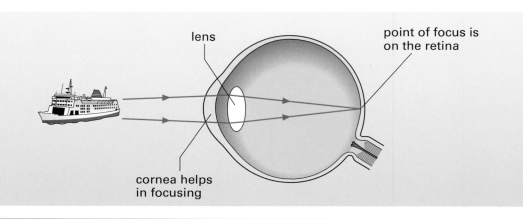

lens

point of focus is on the retina

cornea helps in focusing

Q Which layer of the wall of the eye contains most nerve tissue?

B Bright light, dim light

1 The eye can <u>adapt</u> to dim and bright light conditions.

2 The iris automatically adjusts the pupil to control the amount of light entering the eye. This is an example of a reflex action.

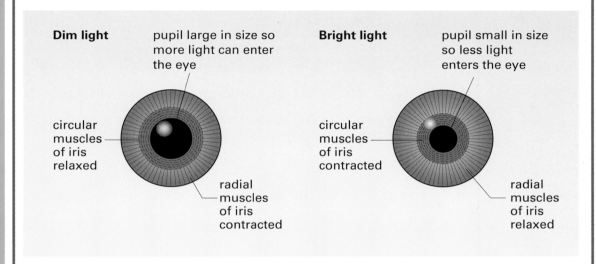

Dim light — pupil large in size so more light can enter the eye

circular muscles of iris relaxed

radial muscles of iris contracted

Bright light — pupil small in size so less light enters the eye

circular muscles of iris contracted

radial muscles of iris relaxed

What path taken by mpulses in the is reflex?

PRACTICE

1 Describe three ways in which the eye is protected from damage.

2 Conjunctivitis is an infection of the membrane covering the outer surface of the eye. What natural mechanism is there to combat infection?

3 Which layer in the wall of the eye is most responsible for transporting food and oxygen?

4 What is the function of the cornea?

5 Why is the optic nerve described as a sensory nerve?

6 The image formed at the point of focus on the retina (see diagram at the bottom of page 34) would appear to be upside down. Why is it that the object looks the right way up to the viewer?

7 If an image is not focused on the retina, how might the object appear to the viewer?

THE BARE BONES

➤ Hormones are part of the chemical coordination system. They help to control conditions within the body and coordinate life processes.

➤ Glands produce hormones, which are carried by the blood.

➤ Hormones act on target organs and tissues to induce a response.

A Glands and hormones

KEY FACT

1 Hormones may <u>act over a period of time</u>.

- Growth hormone adjusts the rate of growth during childhood.
- Sex hormones cause egg production by females during much of their adulthood, as part of the menstrual cycle.
- Insulin helps control sugar level throughout life.

KEY FACT

2 The effects of hormones can also be <u>immediate</u>.

- Adrenaline causes the heart rate to quicken, raises the level of glucose in blood and diverts blood to the muscles and the brain – preparing the body for action.
- Insulin has an immediate effect on blood sugar level, reducing it.

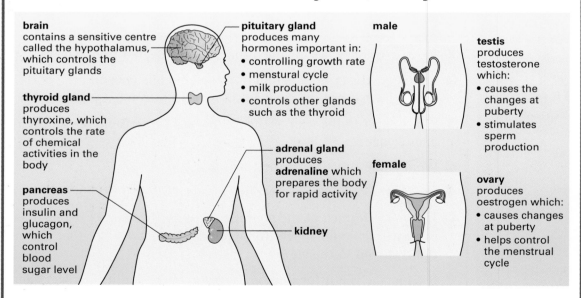

brain contains a sensitive centre called the hypothalamus, which controls the pituitary glands

thyroid gland produces thyroxine, which controls the rate of chemical activities in the body

pancreas produces insulin and glucagon, which control blood sugar level

pituitary gland produces many hormones important in:
- controlling growth rate
- menstural cycle
- milk production
- controls other glands such as the thyroid

adrenal gland produces **adrenaline** which prepares the body for rapid activity

kidney

male

testis produces testosterone which:
- causes the changes at puberty
- stimulates sperm production

female

ovary produces oestrogen which:
- causes changes at puberty
- helps control the menstrual cycle

KEY FACT

3 In sports, it is <u>illegal</u> to use hormones such as <u>testosterone</u> for <u>increasing muscle</u> development.

The side-effects of using testosterone in this way can include:

in males: decrease in sperm production and impotence

in females: increase in body hair, voice deepening and irregular periods.

Q How is the pituitary gland different to other glands?

B Controlling sugar level in blood

1 The <u>amount of sugar</u> in blood is <u>critical</u>, because it has an effect on the nearby cells.

- Too much sugar in the blood draws too much water out of cells, damaging them.
- Low blood sugar is dangerous because it is needed as an energy supply for cells.

2 The blood sugar level has to be <u>adjusted</u> constantly.

- The amount and type of food we eat varies daily.
- The activities we're involved in affect how quickly the food is used by the body.

Remember

he pancreas roduces <u>insulin nd glucagon</u>. he <u>liver</u> is mportant, too, ecause, in the resence of hese hormones, t brings about he changes eeded to ontrol blood ugar level.

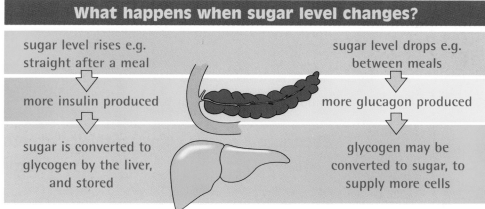

What happens when sugar level changes?

sugar level rises e.g. straight after a meal	sugar level drops e.g. between meals
more insulin produced	more glucagon produced
sugar is converted to glycogen by the liver, and stored	glycogen may be converted to sugar, to supply more cells

Case study

- Someone with diabetes does **not** make **enough insulin** to control the level of glucose in the blood. People with diabetes are called diabetics.
- The most common type of diabetes is inherited. A diabetic lacks the gene that codes for insulin production, so the special cells in the pancreas do not make it.
- Nowadays, insulin is made by bacteria, which have been given the gene for making human insulin. This is an example of a genetically-engineered medicine.
- Less common is type II diabetes. This starts later in life and is linked to obesity. It can usually be controlled by restricting all carbohydrate levels in the diet.

Describe what is meant by a genetically-engineered medicine.

PRACTICE

1 How does insulin (produced by the pancreas) reach the liver, where it mainly acts?

2 Name a target organ for female sex hormones.

3 Describe one effect that hormones cause in males at puberty.

4 Why is the brain important in the hormone system?

5 A friend spent all his money on sweets and ate all of them in one go. What would be the immediate effect on a) blood sugar level? b) insulin production?

6 Suggest two reasons why diabetics have to monitor their diet carefully.

Sexual maturity

THE BARE BONES

➤ Puberty is the time in adolescence when the sex organs mature.
➤ Sex hormones cause the changes during puberty.
➤ The menstrual cycle is controlled by hormones.
➤ Fertility is the likelihood of mating resulting in fertilisation.

A Puberty

KEY FACTS

1 The body changes which happen during puberty are called the <u>secondary sexual characteristics</u>.

2 Secondary sexual characteristics are brought about by <u>sex hormones</u>.

- **testosterone** in males
- **oestrogen** and **progesterone** in females

Remember
Hormones can be used to increase or decrease fertility. It is important to have enough information to make the right decisions about controlling fertility.

Q Can you name and spell the names of the human sex hormones correctly?

Secondary sexual characteristics at puberty:

males	females

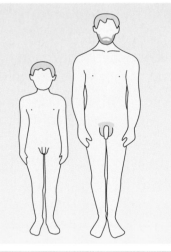 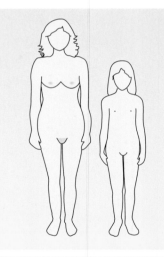

males	females
puberty happens at around 14-16 years	puberty happens at around 11-13 years
whole body has a growth spurt	whole body has a growth spurt
body becomes more muscular	hips widen, buttocks and thighs get fatter
pubic hair grows	pubic hair grows
beard grows	ovaries produce eggs
penis gets larger	periods start
testes produce sperms	

B Menstrual cycle

Having a period is called <u>menstruation</u>. Periods <u>start at puberty</u> and <u>finish at menopause</u> (around 50 years old).

Hormones secreted by the pituitary and the ovaries bring about monthly changes:

- ovulation, when an egg is released from an ovary
- thickening of the womb lining, so that it is ready to receive an egg if is fertilised.

Example: Understanding the menstrual cycle

1 2 3 4 5 6 7 8 9 10 11 12 13 14 15 16 17 18 19 20 21 22 23 24 25 26 27 28	days	
menst-ruation – a period of bleeding	egg develops	ovulation	remains of follicle	What events happen in the ovary during the menstrual cycle?
		How thick is the lining of the womb?
oestrogen	FSH progesterone	What happens to hormone levels?

If a girl has a period from the 10–14th of September, the start of her period is day 1 of the menstrual cycle. She can expect the next period to start 28 days later.

September has 30 days. From the 10th to the 30th September is 20 days, so her next period should start 8 days after that on the 8th October.

C Controlling fertility

1 One reason for low fertility or <u>infertility</u>, is that a woman may not produce <u>ova</u> (eggs). In this case, <u>fertility drugs</u> may <u>stimulate the release of eggs</u> from the ovaries.

2 Contraception reduces fertility by preventing an egg being fertilised by a sperm.

The main ways of doing this are:

prevent sperms from coming into contact with eggs	stopping eggs from being released by the ovaries
preventing fertilised eggs from implanting in the lining of the womb	avoiding ovulation altogether

1 Name the male and female human sex cells.

2 What causes the bleeding during menstruation?

3 How can hormones be used to increase fertility?

4 In some cultures, people think that using an artificial method of contraception is wrong. Which method of contraception might they still consider using?

Body systems for reproducing

THE BARE BONES

➤ The human reproductive systems are designed for the production of sex cells, internal fertilisation and the development of offspring.

➤ There are potential benefits and problems with using reproductive technologies for controlling fertility.

A The female reproductive system

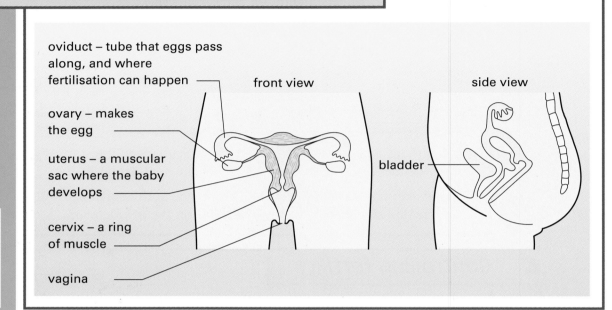

oviduct – tube that eggs pass along, and where fertilisation can happen

ovary – makes the egg

uterus – a muscular sac where the baby develops

cervix – a ring of muscle

vagina

front view

side view

bladder

Q Where are female sex cells:

• produced?
• fertilised?

B The male reproductive system

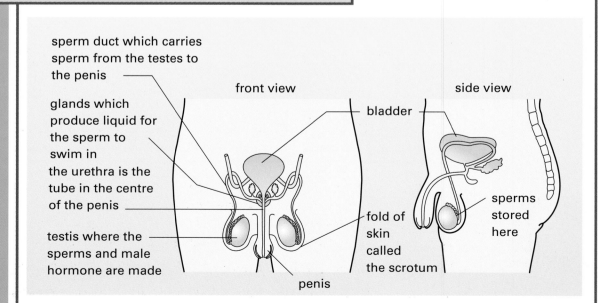

sperm duct which carries sperm from the testes to the penis

glands which produce liquid for the sperm to swim in

the urethra is the tube in the centre of the penis

testis where the sperms and male hormone are made

front view

side view

bladder

fold of skin called the scrotum

sperms stored here

penis

Q Where are sperm cells:

• produced?
• stored?

C *Fertilisation*

1 <u>Fertilisation</u> occurs in the <u>oviduct</u> as the sperm's nucleus enters the ovum, fusing with its nucleus. The DNA in the new nucleus has all the information for the development of a new offspring.

2 The womb is a good place for the foetus to develop because it protects the foetus from injury and is a stable, sheltered environment. The womb can expand greatly to allow growth and it contains the placenta, which supplies the foetus with nutrition and oxygen.

3 *Case study* In vitro fertilisation

There are many technologies to help people prevent fertilisation and increase fertility. In the picture below, people are talking about their experiences using these technologies, and their benefits and problems. Discuss some of the questions with a friend or adult.

Sarah and Tom tried *in vitro* technology to help them start a family. This involves fertilising an egg outside a woman's body and placing it in her womb.

- How might a doctor decide whether or not *in vitro* fertilisation is provided through state health care, or if a couple have to pay themselves?
- Give a reason why Sarah said she felt a failure when *in vitro* fertilisation didn't work.
- How might several years of treatment place a strain on their relationship?
- If the sperm was from another man, how might this affect a couple's relationship?
- What is the argument against people profiting from helping others to start a family?

'We spent a few years and a lot of money trying to have a baby and in the end it didn't work. It was upsetting for me, I felt such a failure.'

'I used to do a lot of overtime to raise extra money, which meant we didn't spend as much time together'

Sharma and esmond use ontraception ecause they on't want to tart a family ow. What do ou think about heir comments?

1 Match the following parts of the female reproductive system with an event that happens there: **ovary, oviduct, uterus, vagina**
a) a secure place for the foetus to develop
b) sperms enter through here
c) eggs may be fertilised as they pass along here
d)   produces eggs.

2 Why could medicines, which are normally safe to use, be unsafe for pregnant women?

3 Why does *in vitro* fertilisation technology not always result in pregnancy?

Keeping the body in balance

THE BARE BONES

➤ Cells work best if the conditions within them stay fairly constant, and this is what is meant by homeostasis.

➤ Balancing the body's needs for water and salts is mainly achieved by the kidneys.

A Homeostasis

KEY FACTS

1 Chemical reactions in cells are controlled by enzymes. Enzymes are sensitive to changing conditions in and around cells, and stop working if the conditions are wrong. Page 17 explains why in more detail.

2 <u>Waste</u> products need to be <u>removed</u> from the body to help keep conditions constant.

- **Carbon dioxide** diffuses from cells into the blood, and from the blood into the air sacs in the lungs, and is breathed out (see page 26).

- **Urea** is waste made when proteins the body does not need are broken down. Urea is mainly excreted by the kidneys in urine, and some is lost in sweat.

3 <u>Defending</u> the body <u>against disease</u> helps to keep body conditions constant. The skin, blood and stomach acid all play a part in defence against disease (see page 24).

Remember
Body conditions do fluctuate, but homeostasis keeps the changes to a minimum.

Q Can you list the main body conditions which need to be controlled?

B Balancing water and salts

KEY FACT

A small amount of <u>water is lost</u> from the body when we <u>breathe out</u>, and some is lost through <u>sweating</u>. Most excess water is filtered into <u>urine</u> by the kidneys.

Q At a party, a child drinks three cola drinks in a row. What would you expect to happen to the amount of urine produced?

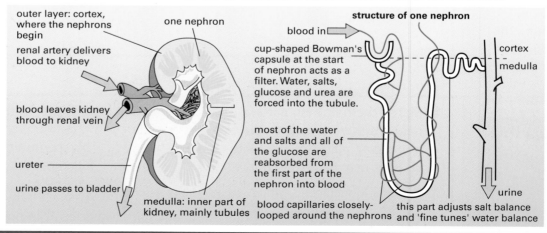

outer layer: cortex, where the nephrons begin

renal artery delivers blood to kidney

one nephron

blood leaves kidney through renal vein

ureter

urine passes to bladder

medulla: inner part of kidney, mainly tubules

structure of one nephron

blood in

cup-shaped Bowman's capsule at the start of nephron acts as a filter. Water, salts, glucose and urea are forced into the tubule.

most of the water and salts and all of the glucose are reabsorbed from the first part of the nephron into blood

blood capillaries closely-looped around the nephrons

cortex
medulla

urine

this part adjusts salt balance and 'fine tunes' water balance

C Temperature regulation

1 Humans and other large animals have some control over body temperature, whatever changes are happening outside. This is called <u>thermoregulation</u>.

- Thermoregulation means that humans can exploit more environments than they could otherwise.

Ways of regulating body temperature	
too cold – warm the body up	**too hot – cool the body down**
<u>shivering</u> - muscle movement which warms the body up	<u>no shivering</u>
<u>sweating decreases</u> so less heat is lost as sweat evaporates	<u>sweating increases</u> – so more heat is lost as sweat evaporates
<u>blood circulation</u> at skin surface <u>decreases</u> so less heat is lost	<u>blood circulation</u> at skin surface <u>increases</u> so more heat is lost
changing behaviour e.g. wrap up warmly, sit by a fire, exercise, have a hot drink	changing behaviour e.g. take off a jumper, sit in the shade, have an iced drink

2 The skin is vital because it:
- helps regulate body temperature by adjusting the blood circulation near the body surface, and the amount of sweat
- is waterproof and protects the body from drying out
- protects against infection
- excretes urea in sweat

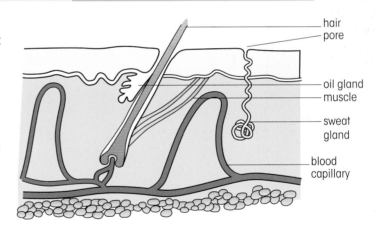

hair
pore
oil gland
muscle
sweat gland
blood capillary

Q Suggest a reason why skin damage over a large surface area (e.g. because of burns) can be fatal.

1 One Christmas Day a child eats the contents of a whole chocolate selection box.
a) Which body condition would need to be controlled?
b) Which body systems are involved in controlling this condition?

2 Why are humans able to live in the Arctic when animals, such as frogs, cannot?

3 How do people help keep their water balance stable when they work out in a gym?

4 Which components of skin are involved in thermoregulation?

Plants and photosynthesis

➤ Plants use photosynthesis to get nutrition. During photosynthesis, plants make carbohydrate food from simple raw materials.

➤ The rate of photosynthesis depends on various factors, including light intensity, availability of raw materials and temperature.

A Photosynthesis

KEY FACTS

1. Photosynthesis takes place in the <u>chloroplasts</u>, where <u>light energy</u> is absorbed by <u>chlorophyll</u>.

2. The <u>energy</u> is needed for <u>converting simple raw materials</u> (carbon dioxide and water) into <u>glucose</u>. <u>Oxygen</u> is released as a <u>by-product</u>.

3. Glucose may be converted to <u>starch</u>, and is stored mainly in the leaves.

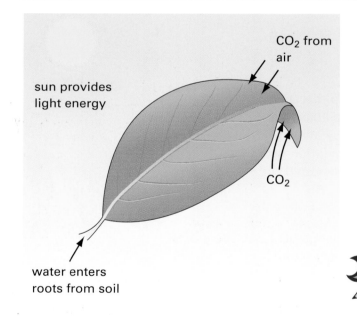

Both photosynthesis and respiration happen in plants.
• Photosynthesis only occurs when there is sufficient light.
  • Respiration occurs all day and night.

Q Why is photosynthesis described as a 'building up' process?

carbon dioxide + water → glucose + oxygen
[+ light energy] (carbohydrate food) (by-product)

KEY FACT

4. Plants use the glucose as a supply of <u>energy</u> and <u>raw materials</u>. These are needed for building up new materials for <u>cell growth</u>.

B Investigating the rate of photosynthesis

1 The rate of photosynthesis depends on several factors, including:

- the availability of light
- the amount of raw materials, e.g. carbon dioxide (CO_2)
- the amount of chlorophyll
- a suitable temperature.

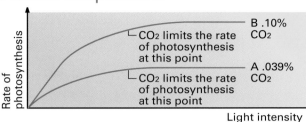

Graph: Rate of photosynthesis vs Light intensity

B .10% CO_2
— CO_2 limits the rate of photosynthesis at this point

A .039% CO_2
— CO_2 limits the rate of photosynthesis at this point

Apparatus diagram:
- beaker
- bubble of oxygen gas
- water
- funnel
- pondweed

Apparatus for investigating rate of photosynthesis, using pond weed.

2 Plant growers can use artificial lights and add extra carbon dioxide to greenhouses to help increase:
- the rate of photosynthesis • how fast plants grow.

Example

If you have to mark out the scale on a graph, try and use the same amount for each axis. In this example 1 cm of graph paper is equivalent to:

- 20 bubbles per min on the y-axis   • 20 cm from the lamp on the x-axis

Draw the best line, which runs through most of the points but may miss one point that does not fit the general pattern.

Read off the graph e.g. 'predict the number of bubbles per minute if the lamp is placed at 70 cm': draw a line from the 70 cm mark on the x-axis to meet the graph line, then go across to the y-axis = 26 bubbles per minute

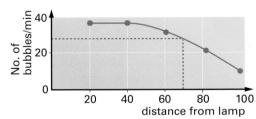

Graph: No. of bubbles/min (0, 20, 40) vs distance from lamp (20, 40, 60, 80, 100)

1 How do the raw materials for photosynthesis reach plant cells?

2 Why might shoot tips, flower buds and root tips be using sugar most actively?

3 How can you tell from the graph at the top of this page that the concentration of CO_2 was a limiting factor at 0.03%?

4 At what time of year might plant growers need to:
a) decrease temperature in a greenhouse?
b) increase lighting?

5 At what distance from the lamp would you expect the plant in the experiment above to produce 34 bubbles of oxygen per minute?

Transport in plants

THE BARE BONES

➤ Transpiration is the loss of water from the leaf surface, which helps to draw water up through a plant.

➤ Minerals are important for healthy plant growth.

➤ Water is important for support, transport and as a raw material.

A Water and plant cells

KEY FACTS

1   Osmosis is the diffusion of water molecules where there is a difference in concentration of water molecules.

2   Water molecules pass from cell to cell, through a surface membrane, along a concentration gradient.

• Water molecules move from an area with more water molecules (a dilute solution) to an area where there are fewer water molecules (a concentrated solution).

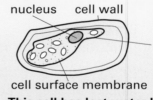

cell wall
semi-permeable cell surface
vacuole
cytoplasm
dilute solution outside the cell
nucleus
chloroplast

3   Water enters plant cells by osmosis, providing rigidity and support. On losing water, the plant cells become floppy.

Q Would the cells of a peeled potato put in salted water gain or lose water?

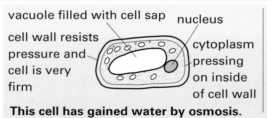

vacuole filled with cell sap
cell wall resists pressure and cell is very firm
nucleus
cytoplasm pressing on inside of cell wall
This cell has gained water by osmosis.

nucleus   cell wall
cytoplasm shrunken and not pressing on cell wall, so the cell is limp
cell surface membrane
This cell has lost water by osmosis.

B Mineral nutrition

KEY FACT

1   Minerals are particles dissolved from rock surfaces. They form a solution with water in the soil.

Minerals are needed for plants to grow healthily. Nitrates are needed for making proteins for good leaf growth. Potassium and phosphate are important for chemical processes, and bud and root growth.

KEY FACT

2   The transport system of plants is made up of two tissues.

Q What is the difference between organic and inorganic fertilisers?

• Minerals are carried in solution in xylem vessels, from the roots to the shoots.
• Dissolved sugar is carried by phloem cells around the plant. Fertilisers contain minerals. Organic fertilisers (manure, compost) are made from the remains of living things. Inorganic fertilisers (which often contain phosphate, nitrogen and potassium) are made from chemicals in factories.

c   Transpiration

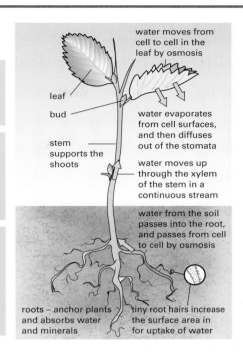

water moves from cell to cell in the leaf by osmosis

leaf

bud

stem supports the shoots

water evaporates from cell surfaces, and then diffuses out of the stomata

water moves up through the xylem of the stem in a continuous stream

water from the soil passes into the root, and passes from cell to cell by osmosis

roots – anchor plants and absorbs water and minerals

tiny root hairs increase the surface area in for uptake of water

1 Cell surfaces are moist. Water <u>evaporates</u> from them into the spaces between.

2 During <u>transpiration</u>, water vapour <u>diffuses out</u> of pores, called <u>stomata</u>, between the cells at the leaf's surface.

3 Leaf cells replace lost water from the solution inside <u>xylem vessels</u>, tiny tubes running through the plant from the root to the shoot.

Remember
Plants need stomata because they have to exchange gases with their surroundings. As a result, water vapour is lost from leaves too.

What conditions affect the rate of transpiration?	
condition	reason it affects transpiration
temperature increases	water molecules evaporate from cell surfaces more quickly
a windy day	moving air takes water vapour away from the leaf surface, increasing the diffusion gradient
dry air around the leaf	water evaporates from cells faster
more stomata (mainly on the lower surface)	faster transpiration
waterproof waxy cuticle (usually on upper surface)	prevents water moving through it
amount of light (because stomata close in the dark)	no gas exchange occurs

Q What conditions slow down the rate of transpiration?

1 Why does a plant lose water from its leaves?

2 How does water move along a concentration gradient?

3 How does water move out of the stomata on the leaf surface?

4 A student took two leaves of the same type and size. He covered the lower side of leaf A with grease, and the upper side of leaf B with grease. He weighed both leaves and left them by a sunny window. He reweighed them 4 hours later. Here are the results: *Leaf A lost 2% of its mass, and leaf B lost 10% of its mass.* Explain the difference in amount of water lost by leaf A and leaf B.

Controlling plant growth

THE BARE BONES

➤ Plants make growth movements which are called tropisms.
➤ Hormones control how plants grow.
➤ Hormones are used in horticulture and agriculture.

A Control of plant growth

KEY FACT

1 Plants respond to <u>light</u>, <u>moisture</u> and <u>gravity</u> by growing in a particular direction.

• shoots grow towards light and away from gravity
• roots grow towards moisture and gravity

The advantage is that whichever way a seed lands in the soil, its shoot will grow upwards and its root will grow downwards.

KEY FACT

2 <u>Hormones</u> are chemicals that <u>coordinate plant growth</u>.

Hormones work at **low concentrations**. They influence how fast cells divide and the elongation of cells at the tip of roots and shoots, speeding it up or slowing it down.

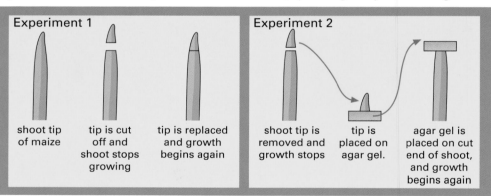

Experiment 1

shoot tip of maize / tip is cut off and shoot stops growing / tip is replaced and growth begins again

Experiment 2

shoot tip is removed and growth stops / tip is placed on agar gel. / agar gel is placed on cut end of shoot, and growth begins again

Q How do hormones help plants get maximum light in a shady wood?

3 Hormones can cause uneven growth of shoot or root tips, so that one side grows more quickly.

higher auxin concentration in the root slows growth

higher auxin concentration in the shoot increases growth

• Auxin is a hormone which increases cell growth rate in shoots.
• Auxin slows cell growth rate in roots.

KEY FACT

4 <u>Phototropism</u> is the growth of a plant towards light.

• Auxin is involved in phototropism.

• There is more auxin on the side of the shoot away from light, so it grows faster and curves towards light.

B Plant science in business

1 Auxins can be used to:
- generate many plants from cuttings • kill weeds
- regulate fruit ripening during transport.

Auxin experiments		
procedure	effect	how it works
cut the bud off the main shoot	more side shoots develop	auxin from main bud inhibits growth of side shoots
spray fruit with auxin	more fruit sets	increases pollination, fertilisation and fruit formation
dip shoot cuttings in auxin-containing powder	roots grow faster from the cut shoot	auxin stimulates the growth of small roots (although it can inhibit larger side roots)
spray auxin-containing weed killer	weeds die	auxin disturbs cell chemistry and causes excessive growth

Auxin weed killers are absorbed more rapidly by broad-leaved plants than grasses. Why does this make them suitable for use on lawns?

2 Ethene is produced in low concentrations by plants, and causes ripening of fruits. Unripe fruits are transported because they are less likely to be damaged or rot. Just before arrival at supermarkets, ethene is used to bring on ripening.

PRACTICE

1 How does gravity affect plant shoots and roots?

2 What is auxin? How might scientists find out if there is auxin in a shoot tip?

3 How does auxin from the tip of a shoot affect cell growth in that shoot?

4 What effect does auxin from a main shoot have on side shoots?

5 Suggest why putting a ripe fruit with unripe ones tend to speed up ripening.

Variation and genetics

THE BARE BONES
➤ Variation is caused genetically or by factors in the environment.
➤ DNA is the genetic material which acts as a code for inherited characteristics.

A What is variation

KEY FACTS

1 Variation describes the differences between living things. For example, think about differences such as eye, hair and skin colour between people you know.

2 Genetic variation is caused by genes which are inherited by offspring from their parents. This:

- mainly occurs when the genetic material divides up during cell division, to form sex cells.

- sometimes occurs suddenly, due to a chance change called a mutation.

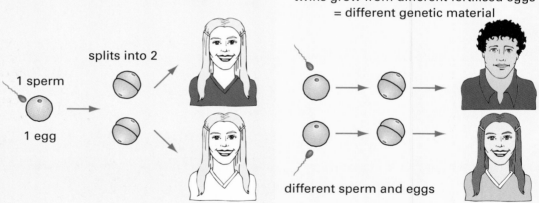

Identical twins = same genetic material

1 sperm
splits into 2
1 egg

Brothers and sisters including non-identical twins grow from different fertilised eggs = different genetic material

different sperm and eggs

3 Environmental factors can affect the characteristics of living things e.g.:

diet and nutrition   light   disease   lifestyle factors

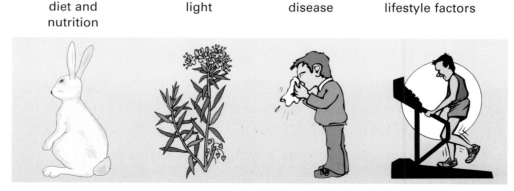

Q Dalmatian puppies may be very similar in appearance but they do not all have spots in the same pattern. How is this variation caused?

B About genetic material

1 DNA makes up most of the <u>nucleus</u> of a cell. The pattern of particles making up DNA acts as a <u>code</u> or set of instructions for cell processes.

2 DNA is organised into <u>chromosomes</u>, which are:

- always in **pairs in body cells** e.g. muscle or skin cell. Each human cell has 23 pairs of chromosomes in the nucleus
- always **single in sex cells** (called gametes) within sex organs. Human sex cells (sperm and ovum) have 23 single chromosomes
- made up of genes, which are organised along the chromosomes and code for different proteins.

3 A gene is a short chunk of DNA which codes for a particular protein. Proteins give rise to particular characteristics.

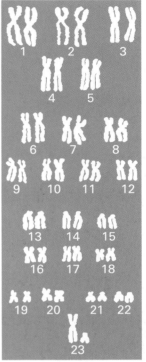

The 23 pairs of chromosomes, found in humans

Example

protein which a gene codes for	inherited characteristic
insulin	ability to control sugar level in blood
blood clotting factor	blood clots at the site of an injury

In many cases, a characteristic is caused by a combination of more than one gene.

4 The human genome project was carried out by scientists worldwide to map the exact position and function of all the genes on each chromosome. This helps us to develop gene therapies, for genetic diseases e.g. cystic fibrosis.

What is DNA? You need to be sure of its function and how it is organised.

1 How does variation happen?

2 How can variation lead to both extinction and evolution?

Evolution and selection

THE BARE BONES

➤ Although people have different views about inheritance and evolution, scientists accept the main ideas of Darwin and Wallace.

➤ The main ways of influencing inherited characteristics are cloning, selective breeding and genetic engineering.

A Creation and evolution

KEY FACT

1 A few hundred years ago, it was commonly believed that the Earth and living things were spontaneously created by God, as described in the Bible. This <u>theory of creation</u> implies that:

- all life was present at the start of the Earth
- life forms have not changed since the beginning

KEY FACT

2 In 1809, Jean Baptiste de Lamarck proposed that life evolves because of <u>changes</u> that happen to animals <u>during their lifetime</u>. They <u>acquire,</u> or gain, characteristics and can <u>pass them on</u>.

Example: People who do physical work develop strong muscles, and so their offspring would have strong muscles too.

Lamarck's ideas were not accepted because people believed in creation, but it made people think about evolution.

3 Charles Darwin and Alfred Wallace both contributed to the theory of evolution by <u>natural selection</u>. Charles Darwin based his ideas on years of study of many animal species and fossils, carried out while charting maps in the southern hemisphere.

KEY FACT

1800
Lamarck

Darwin

1900

2000

- The main points are that mutations cause genetic variation, and give rise to new characteristics
- a new characteristic may be beneficial or harmful to the survival of an individual
- a beneficial characteristic is more likely to be passed on to future generations

Darwin's ideas only became accepted after his death, as people's views changed and scientists provided more evidence.

Q Why were Lamarck's ideas important even though they were not accepted fully?

B Selective breeding

1 Selective breeding means choosing only plants and animals with desirable characteristics to breed from.

2 Selective breeding is:

- a long term project because it can take many generations to breed offspring with the right combination of characteristics
- unpredictable, because the exact combination of inherited characteristics is not know until the offspring is born.

3 Selective breeding is very important in agriculture, because it has resulted in increased yields from crops, plants and farm animals.

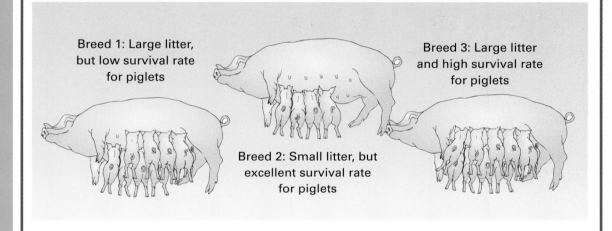

Breed 1: Large litter, but low survival rate for piglets

Breed 2: Small litter, but excellent survival rate for piglets

Breed 3: Large litter and high survival rate for piglets

Suggest an advantage and a disadvantage of using selective breeding in farming.

1 A particular type of moth exists in dark and light forms. In an experiment some years ago, moths were released in different areas of the UK and recaptured later. The table shows data.

area of UK	experiment	number of light - coloured moths	number of dark - coloured moths
unpolluted	released	488	485
	recaptured	59	30
polluted with soot	released	62	157
	recaptured	15	83

a) What percentage of dark-coloured moths were recaptured in the two areas?
b) What factors might affect the survival rate of dark-coloured moths in each area?
c) Predict the effect of reducing pollution on both types of moth.

2 Suggest characteristics that farmers might find useful in selective breeding of:
a) cattle for milk production
b) pigs for meat production
c) wheat for fast cropping.

THE BARE BONES

➤ Living things reproduce in two main ways: asexually and sexually.
➤ Genetic engineering is a way of changing inherited characteristics.
➤ Cloning is a type of asexual reproduction.
➤ Genetic engineering has been used to produce better medicines.

A   Asexual reproduction

KEY FACTS

1 Asexual reproduction involves only one parent. The process involves an increase in cell numbers, and all the new cells are identical genetically.

• There is **no variation**.   • The identical offspring are called **clones**.

2 Examples of asexual reproduction include producing runners or bulbs, and taking shoot cuttings.

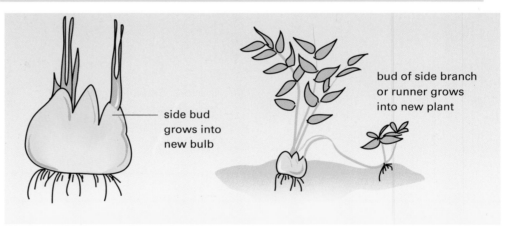

side bud grows into new bulb

bud of side branch or runner grows into new plant

Case study

Cloning

Plant cloning is reliable because the numerous offspring have the same features as the parent plant. It is common and important commercially.

One method is taking shoot cuttings, and encouraging them to root and grow into adult plants e.g. geraniums. Another method is taking a small lump of tissue which contains cells that have not yet specialised, and growing them in a culture medium. Orchids, oil palms and cauliflowers are cloned in this way.

Cloning animals involves splitting the early embryo of a fertilised egg and putting each half into another egg cell. The egg cell is then placed in the womb of an adult female to develop.

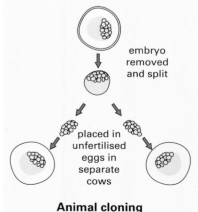

embryo removed and split

placed in unfertilised eggs in separate cows

Animal cloning

Q A gardener takes a potato and cuts it into three pieces, each with a bud that will grow into a new plant. Explain why this is an example of asexual reproduction.

B   Sexual reproduction

> Fertilisation happens during <u>sexual reproduction</u>, when the nucleus of one sex cell fuses with the nucleus of another.

1 The parents produce sex cells which are genetically different.

2 The offspring resemble their parents because they inherit their characteristics from them.

3 The offspring inherit a different combination of chromosomes to each other and the parents, creating genetic variation.

4 Variation allows the chance of stronger characteristics arising in a population.

Why are brothers and sisters similar but not the same?

C   Genetic engineering

> **1** Genetic engineering involves transferring genes artificially from one organism to another.

Example: There is a human gene which codes for insulin production.

- the gene is 'cut out' using enzymes, and transferred to a bacterium
- bacteria are cultured and produce insulin on a large scale.

Advantage = insulin is pure, safe for use in humans and available cheaply for people who are diabetic

How might genetic engineering influence evolution?

> **2** There is some concern about genetic engineering because genetically engineered organisms may escape into the natural environment, and we do not know what the effects might be.

PRACTICE

1 Which of the following are asexual reproduction? Which are sexual reproduction?

   a) You dig up a clump of lilies, divide it into smaller clumps and replant several.
   b) Frog spawn appears in the garden pond each Spring.
   c) A vet fertilises a cow using sperm from the farmer's prize bull.
   d) Berries develop on a holly tree each year.

2 GM foods contain ingredients from genetically-engineered plants. Why might some people not want to eat GM foods?

3 How might genetic engineering be used to help make a gene therapy for treating people with an inherited disease?

THE BARE BONES

➤ Separately inherited factors called genes are responsible for the inheritance of characteristics.

➤ It is possible to predict simple patterns of inheritance, using genetic diagrams.

A Gregor Mendel and inheritance

1 Gregor Mendel discovered the basics of how inheritance happens. He carried out thousands of crosses between pea plants, studying seven different characteristics including flower colour, stem length, colour of seed coat.

2 Mendel started with plants that had only produced red or white flowers for several generations, so only contained one type of factor or allele.

3 He crossed red and white flowered plants, took seeds and grew the offspring and then crossed these plants again, for several generations.

4 All flowers were red or white (no pink flowers occurred).

5 White flower colour disappeared in the first generation, but reappeared in the next.

KEY FACT

Q Which piece of evidence made Mendel think that factors (alleles) do not mix, but are inherited separately?

B Predicting inheritance

1 A gene codes for a particular characteristic, such as eye colour. Alleles are different forms of the same gene e.g. an allele for blue eye colour or an allele for brown eye colour.

2 Offspring inherit one allele from each parent for a characteristic.

• A dominant allele will hide the appearance of a recessive allele, even if there is only one of them, e.g. brown eye colour is dominant over blue eye colour.

• Write a dominant allele with a capital letter and a recessive allele with a lower case letter, eg: B = brown eye colour, b= blue eye colour

Example: Inheriting freckles

The allele for freckles (F) is dominant to the allele for no freckles (f). Write the alleles of the parents (female Ff, male Ff) alongside the crossing diagram and then fill in the possible combinations for each square. Use the outcomes to decide how each offspring will look:
ff = no freckles Ff and FF = freckles.

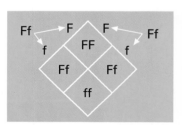

KEY FACTS

Q Imagine a child inherits these alleles:
• Bb
• FF
• XY
(see page 57).
What characteristics will the child have?

B

In humans, one pair of chromosomes controls gender, XX in females and XY in males. Around 50% of human births are female and 50% are male.

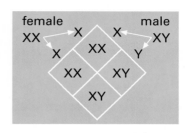

C   Inheriting diseases

1   Inherited diseases are often recessive. Carriers of the allele may not show symptoms, as the normal gene is dominant.

2 Cystic fibrosis and haemophilia are examples of recessive inherited diseases.

3 Huntingdon's disease is a disorder of the nervous system that can be passed on by just one parent, because it is caused by a dominant allele.

4 Sickle-cell anaemia is caused by a recessive allele, resulting in abnormally-shaped haemoglobin and red blood cells. Carrying one recessive allele can help protect people against malaria

5 Case study: Cystic fibrosis

Around 50% of Europeans carry one allele for cystic fibrosis, but suffer no effects as it is a recessive allele. Someone with cystic fibrosis produces sticky mucus in the lungs and gut. The lungs get blocked up easily and fluid collects, and there are problems with digestion.

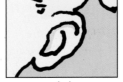

Which examples of genetic diseases do you need to know for your exam board?

How likely is t that two eople will have child with ystic fibrosis, if heir alleles are c and CC where c = ystic fibrosis nd C = ormal)?

PRACTICE

1 In Mendel's experiments, which piece of evidence suggests that one allele can be dominant over another?

2 Taking R to represent the allele for red flower and r to represent the allele for white flower, draw out a cross between a red flower and white flower as shown in Example 1. What colour flowers would all the offspring have?

3 Use a genetic diagram (as shown in section B) to show what happens when two red plants with mixed alleles (Rr) are crossed. What is the ratio of red and white flowered plants in the offspring?

4 In humans, some people have different ear lobe shapes. Normal ear lobes (D) is dominant to attached earlobes (d)

What are the chances of two people having children with attached ear lobes, if their alleles are dd and Dd?

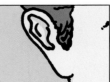

ear lobes droop (called 'free')   ear lobes slope (called 'attached')

Humans and the environment

THE BARE BONES
- ➤ The Earth is a fragile ecosystem, and resources are finite.
- ➤ Humans have an impact on the environment.
- ➤ Only careful management of the environment will safeguard the future of the planet.

A Ecosystems

KEY FACT

1 An ecosystem is made up of living things, the environment they live in and the interactions between them. The Earth is the largest ecosystem, but the climate and shape of the land varies around the world, giving several major ecosystems within it.

Coniferous forest
e.g. Canada
- cold to mild
- acidic soil
- birds and mammals

Temperate forest
e.g. Northern Europe
- mild (no extremes)
- large trees, birds and mammals

Tropical rainforest
e.g. Brazil
- warm, very wet
- huge bio-diversity, rapid plant growth

Desert
e.g. Sahara
- hot daytime, cold night-time
- dry
- sparse vegetation

Tropical grassland
e.g. Central Africa
- hot and dry
- some regular rainfall
- grazing animals and their prey, invertebrates

Tundra
e.g. Siberia
- extreme cold
- water locked as ice
- migrating birds/animals

Q Which of the ecosystems shown in the pictures are more common in Europe?

KEY FACT

2 Each ecosystem supports a different variety of living things. This is called biodiversity.

- Biodiversity is important because it represents the total pool of genes, *and* the potential for new ones.
- Humans get many resources from living things, including medicines.
- Biodiversity is reduced by the environmental impact of many human activities.

KEY FACT

3 Within the ecosystem, living things compete for factors such as food, space, minerals and light. Predation and disease can also affect population size.

B Human impacts on the environment

1 All living things have an <u>impact</u> on the environment. When the human population was small, their impact on the environment was small and local.

2 Humans **exploit resources faster** than other species, especially **non-renewable** resources.

3 The human **population is growing** very fast and the way we live has changed.

• *agriculture* – intensive farming requires the use of agrochemicals e.g. inorganic fertilisers and pesticides and recently, genetic engineering.

• *industry* – getting the raw materials, processing them and distributing products all generate pollution.

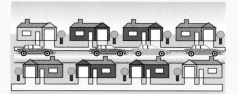

• *the growth of cities* – requiring space, and generating pollution due to increased traffic and domestic heating

• *drastic changes to the ecosystem* – such as deforestation, draining marshland, mining, damming and flooding natural areas, or redirecting waterways.

4 Many human activities cause pollution, which is one reason why they have an undesirable impact on the environment. Long-term pollution, such as heavy metals in ground water supplies, or radiation:

- can have unpredicatable effects
- are not easy or always possible to reverse.

What is the main reason for the increase of human impact on the environment?

1 What are the main components of an ecosystem?

2 Suggest one reason why it is important to maintain biodiversity.

3 Why does burning fossil fuels cause pollution?

4 Giant pandas feed on bamboo. As natural environments are lost, the supply of bamboo is decreasing. Suggest the long-term effect of this on the population of giant pandas, and how the species might be protected from extinction.

THE BARE BONES

➤ We rely on the Earth's resources for all our basic needs of life.
➤ Some of the Earth's resources cannot be renewed.
➤ Human impact on the environment can be reduced by sensible management.

A Renewable and non-renewable resources

KEY FACT

1 Many of the Earth's resources are non-renewable, and we must make important choices about how we use them. Often there are <u>competing priorities</u> because different things are important to different people.

Remember
Sustainable management means using resources in a way that ensures they will be available in the future.

Case study
Fishing for food?

One of the Earth's richest resources is the sea. In the past, people fished successfully in the North Sea and around the Great Banks off northern Canada. In recent years, improved fishing methods meant bigger catches, especially of cod. Despite controls, fish stocks dropped to very low levels and now there is a ban on cod fishing. Scientists hope this will give remaining cod time to grow and reproduce, replenishing stocks.

Alternative strategies might be:

* fish farming e.g. salmon and trout.
* growing other protein-rich foods e.g. pulses and beans, as an alternative food source.

KEY FACT

2 There are many strategies for protecting the future of our planet.

* **Recycling materials**, such as glass, aluminium and paper helps reduce waste and the need for new materials. Local recycling schemes encourage the separation of waste by each household, so that more materials can be re-used.

* **Conservation programmes** help protect sensitive ecosystems e.g. Sites of Special Scientific Interest (SSSIs), where human impacts, such as farming, are restricted.

* **Special programmes** aimed at protecting endangered species which face extinction, e.g. Convention on International Trade in Endangered Species (CITES); captive breeding; national parks and seed banks.

* **Managing ecosystems**, e.g. felling only a small proportion of trees in a forest on a rotation basis, rather than a whole forest area; growing specialist crop plants in small areas within a rain forest, rather than clearing the whole area for farming.

* **Education**, aimed at influencing the attitudes of future generations.

Q Why might these people be interested in fishing management
a) a fisherman?
b) a shopper?
c) a scientist ?
d) a politician?
e) a farmer?

B   The carbon cycle

1   <u>Carbon</u> makes up a large proportion of living cells. It is present in:

- biomass (the bodies of living things) and their remains, including fossil fuels
- the air as carbon dioxide
- the rocks which contain carbonate.

2   <u>Carbon dioxide</u> is <u>removed</u> from the atmosphere by plants, during <u>photosynthesis</u>.

3   <u>Carbon dioxide</u> is <u>added</u> to the atmosphere by living things when:

- they carry out respiration
- things decay
- we burn fossil fuels or other materials from living things e.g. wood

Questions often ask what happens if one factor in an ecosystem changes. Think about the living things affected at once e.g. if foxes increase, rabbits decrease. But also think about further links e.g. fewer rabbits may mean less crop damage.

Which process is carbon dioxide a raw material for?

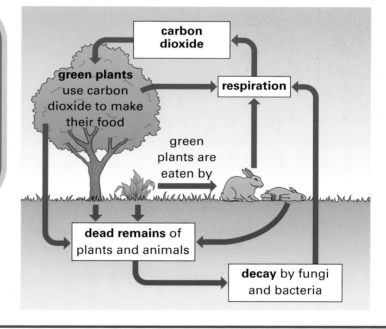

1 Which of the following are renewable and which are non-renewable resources: timber, aluminium, cotton, nylon, paper, North Sea gas, polythene, wool, soya protein?

2 In the long term, why is a ban on cod fishing part of sustainable management?

3 Suggest a species of animal protected by CITES.

4 Which parts of the carbon cycle are most influenced by humans?

5 Why is carbon dioxide an important gas in our atmosphere?

Survival

THE BARE 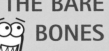 BONES
➤ There are various basic needs of life.
➤ There is competition and interdependence between living things.
➤ Nature recycles resources such as carbon and nitrogen.

A Interdependence in an ecosystem

KEY FACT

1 All the basic needs of life, such as food and minerals, the availability of light and shelter come from the environment. This means that resources are <u>finite</u> – there is only so much of any resource. Living things compete for these resources.

- Some resources, such as minerals, are **naturally recycled** when living things die.
- Light is a **non-renewable** energy source, but unlikely to 'run out' in the imaginable future.

KEY FACTS

2 <u>Survival rate</u> is influenced by the <u>interdependence</u> of living things, <u>competition</u> between them for the basic needs of life, and their <u>adaptation</u> to the environment.

3 Living things are linked through their need for food. <u>Food chains</u> describe the feeding relationships between them (see page 64).

Example: wheat crop → rabbit → fox → invertebrates → microbes

- **Photosynthetic** organisms – mainly plants – are always **at the start** of a food chain, because they are food **producers**.
- Animals are consumers, because they eat ready-made food in the form of plants or other animals.
- Many food chains overlap, forming a food web.

Remember
In a food chain the arrows always lead <u>from the food to the feeder</u>.

KEY FACT

4 A change to the population size of one species, will influence the populations of other living things that feed on them.

A

5 A <u>predator</u> is an animal that eats another, and the <u>prey</u> is the animal which is eaten. The sizes of the two populations are <u>interdependent</u>

Case study: Predator and prey

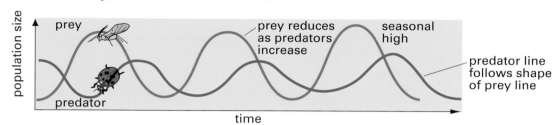

The populations in the graph above show a typical shape: as each population rises and falls, the shape of the predator graph follows the shape of the prey graph. There may also be a seasonal effect, for example, a hard winter may reduce the number of ladybirds the following spring, so the greenfly population increases faster than the year before. Biological control uses natural predators to control a pest population, which is kept at low levels but not eradicated, e.g. parasitic wasps control aphid infestations.

Name one roducer shown n this graph, nd pick out ne food chain.

B Adaptations and survival

Living things are adapted differently to survive in a range of ecosystems, and this gives them an advantage over others that are not.

Adaptation	Arctic fox	Desert rat
body size and surface area	large body size, but smaller surface area compared to body size = less heat loss	small body size, but larger surface area compared to body size = more heat loss
thickness of insulating coat	thick coat	thin coat
amount of body fat	thick layer	thin layer
camouflage	white coat	brown coat

The desert t produces ttle or no rine. How ight this help survive?

1 Imagine that pollution kills most of the water beetle population. What would happen to the other organisms in the food chains that include water beetles? Explain the 'knock-on' effects.

2 Pick out two food chains from the pond graphic. Do they overlap?

3 Large cane toads were introduced into Australia from South America to control rats in sugar cane plantations. The toads are big killers and eat a variety of other prey. Suggest undesirable effects on the ecosystem of introducing this predator.

4 Grasses adapted to very dry conditions tend to have few stomata. Explain why.

THE BARE BONES

➤ Food chains can be described using pictures called pyramids of biomass.
➤ There is energy flow from the Sun, through the levels of a food chain.

A Food chain basics

1 The <u>Sun</u> is the only <u>source of energy</u> for all living things. The <u>energy</u> is temporarily <u>stored</u> in <u>biomass</u>.

2 The stages of a food chain are called <u>trophic</u> levels. Plants (and bacteria which can photosynthesise) are <u>producers</u> and are always the first trophic level.

3 Animals are <u>consumers</u>, eating plants and other animals.

- The second trophic level includes animals called **primary consumers**, which are **herbivores** (mainly eat plants).

- The third trophic level includes animals called **secondary consumers**, which are **carnivores** (mainly eat animals).

Q What is the maximum number of trophic levels shown in the food web on the right?

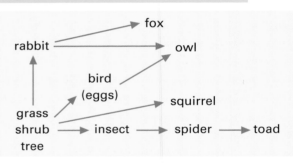

B Food webs as pyramids

The mass of living material at each stage in a food chain can be <u>drawn to scale</u> and shown as a <u>pyramid of biomass</u>.

Here the producers include oak trees, bramble, grass, bluebells, other shrubs and smaller plants. Primary consumers include worms, rabbits, field mice etc. Other consumers include stoat, birds, fox, badger etc.

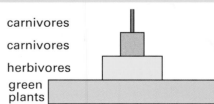

Q Why does the amount of biomass of each trophic level decrease in the food chain?

Data for an English woodland	
Trophic level	g/m²
1 (producers)	6000
2 (primary consumers)	3
3 (secondary consumers)	1.75

1 Decide on a scale e.g. 1 kg = 1 cm on the bar representing the trophic level.

2 Convert data in kg to cms e.g.:
   500 g = 1 cm
   6000 g = 6000 ÷ 500 = 12 cms.

3 Draw to scale, with producers at the base.

c   Food production and energy flow

1 At each stage in a food web, <u>less material and less energy are stored</u> in biomass. This is because some is:
- lost in <u>waste materials</u> • used in <u>repair</u> of cells
- lost as heat during <u>respiration</u>.

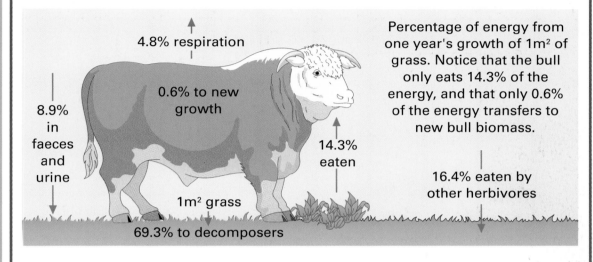

4.8% respiration

0.6% to new growth

8.9% in faeces and urine

14.3% eaten

1m² grass

69.3% to decomposers

Percentage of energy from one year's growth of 1m² of grass. Notice that the bull only eats 14.3% of the energy, and that only 0.6% of the energy transfers to new bull biomass.

16.4% eaten by other herbivores

Q. Which life processes are involved in the flow of energy through a food chain?

2 The amount of energy, flowing through a food chain depends on the plants and animals. In terms of food production, the <u>shorter</u> the food chain, the less energy is wasted.

1 Study this food web:
a) Complete a food chain, starting with grass and finishing with foxes.
b) Using examples from this web, what is meant by a producer and primary consumer?
c) Name two predators in this web.

foxes   owls   thrushes

rabbits   mice   slugs   caterpillars

grass

trophic level	kJ/m²/year
1	88 000
2	14 000
3	1 600
4	100

2 Draw a pyramid of energy using the same method as for a pyramid of biomass, using this energy data taken from a river in USA.

THE BARE BONES

➤ Atoms contain protons, neutrons and electrons.

➤ The atomic number = number of protons or electrons in an atom.

➤ Electrons are arranged in shells (energy levels) around the nucleus.

➤ Isotopes are the same element with different numbers of neutrons.

A Structure of the atom

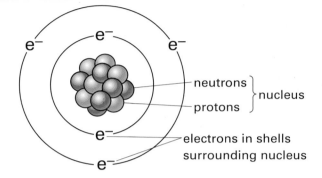

KEY FACT

Atoms have a nucleus containing <u>protons</u> and <u>neutrons</u>, the <u>electrons</u> move around the nucleus.

Protons have a **charge of +1** and a relative mass of 1.

Neutrons have **no charge** and a relative mass of 1.

Electrons have a **charge of −1** and a negligible mass.

Q What is the lightest part of the atom?

B Atomic number and mass number

KEY FACTS

1 The <u>atomic number</u> of an element tells you the number of protons or electrons in an atom.

2 The <u>mass number</u> tells you the number of protons plus the number of neutrons in the atom.

Number of neutrons = mass number – atomic number

How do you work out the number of neutrons in an atom? Take an atom of iron:

mass number ——— ^{56}
atomic number ——— $_{26}$ Fe

Answer

Subtract the atomic number from the mass number (56 – 26 = 30). There are 30 neutrons in an atom of $^{56}_{26}$Fe

Q How many protons and neutrons in a fluorine atom $^{19}_{9}$F?

C Electron arrangement in atoms

1 The periodic table is arranged by electron arrangement. Learn the arrangement for the first 20 elements.

2 Electrons are arranged in <u>shells</u> or <u>energy levels</u> around the nucleus.

• The 1st shell can hold **2 electrons**.

• The 2nd and 3rd shells can hold **8 electrons**.

To work out the electron arrangement for an atom, count the **number of rows above** that element in the periodic table. This is the **number of complete shells** of electrons. Then the number of electrons in the **outer shell** is the same as the **group number** of the element. To check, add up the number of electrons you think you should have in all the shells and this should equal the atomic number. This is useful for working out the arrangement for the first 20 elements.

Example $_{11}$Na, the electron arrangement for sodium is 2.8.1 (2 + 8 + 1 = 11).

Q What is the electron arrangement for carbon $_6$C?

D Isotopes

1 Isotopes are atoms of the same element but with a different number of neutrons.

• They have the **same number of protons**.

• Their **atomic number is the same**, their **mass number is different**.

2 Example

Chlorine has two isotopes:

$^{35}_{17}$Cl and $^{37}_{17}$Cl

$^{35}_{17}$Cl has 18 neutrons and $^{37}_{17}$Cl has 20 neutrons.

Know what is different and what is the same about isotopes of an element.

There are 4 similarities (same number of protons, same number of electrons, same atomic numbers, same symbol) and 3 differences (different numbers of neutrons, different mass numbers).

There is often a question where you will have to work out the number of protons and neutrons in an atom of an element. Make sure you know how to work them out.

Q What part of an atom has a different number in the isotopes of an element?

PRACTICE

1 What parts of the atom have the same mass?

2 How many neutrons are there in a hydrogen atom $^1_1$H?

3 Name two similarities in isotopes of the same element.

The periodic table

THE BARE BONES

➤ The periodic table contains all the known elements.
➤ Groups of elements go down the table.
➤ Periods of elements go across the table.

A How the periodic table is arranged

KEY FACTS

1 <u>The elements are arranged in order of their atomic number.</u>

The metals are found on the left-hand side of the periodic table and the non-metals on the right-hand side of the of the periodic table.

KEY FACT

2 Groups go down the table. They are numbered 1 to 8 in roman numerals (I to VIII).

The group an element is in depends on the number of electrons the element has in its outer shell. It is the number of electrons in the outer shell that largely explains the reactivity of an atom. This is why elements with the same number of electrons in the outer shell – elements in the same group – act similarly in reactions.

KEY FACT

3 The periods go across the periodic table. There are seven periods. The period an element is in depends on which electron shell is filling up.

For instance, there are two elements in the first period as there is only space for two electrons in the 1st shell.

There are 8 elements in the 2nd and 3rd periods as there is room for 8 electrons in each of these shells.

4 From the periodic table, it is possible to predict how reactive an element is:

- for metals, group I is more reactive than group II
- the further down the group the metal is, the more reactive is it
- for non-metals, group VII is more reactive than group VI
- the further up the group the non-metal is, the less reactive it is.

Q Chlorine has 7 electrons in its outer shell, what group will it be in?

B The periodic table

Are there more metals or non-metal in the periodic table?

I	II						transition metals						III	IV	V	VI	VII	O
																		He Helium 2
Li Lithium 3	Be Beryllium 4												B Boron 5	C Carbon 6	N Nitrogen 7	O Oxygen 8	F Fluorine 9	Ne Neon 10
Na Sodium 11	Mg Magnesium 12												Al Aluminium 13	Si Silicon 14	P Phosphorus 15	S Sulphur 16	Cl Chlorine 17	Ar Argon 18
K Potassium 19	Ca Calcium 20	Sc Scandium 21	Ti Titanium 22	V Vanadium 23	Cr Chromium 24	Mn Manganese 25	Fe Iron 26	Co Cobalt 27	Ni Nickel 28	Cu Copper 29	Zn Zinc 30		Ga Gallium 31	Ge Germanium 32	As Arsenic 33	Se Selenium 34	Br Bromine 35	Kr Krypton 36
Rb Rubidium 37	Sr Strontium 38	Y Yttrium 39	Zr Zirconium 40	Nb Niobium 41	Mo Molybdenum 42	Tc Technetium 43	Ru Ruthenium 44	Rh Rhodium 45	Pd Palladium 46	Ag Silver 47	Cd Cadmium 48		In Indium 49	Sn Tin 50	Sb Antimony 51	Te Tellurium 52	I Iodine 53	Xe Xenon 54
Cs Caesium 55	Ba Barium 56	La Lanthanum 57 x	Hf Hafnium 72	Ta Tantalum 73	W Tungsten 74	Re Rhenium 75	Os Osmium 76	Ir Iridium 77	Pt Platinum 78	Au Gold 79	Hg Mercury 80		Tl Thallium 81	Pb Lead 82	Bi Bismuth 83	Po Polonium 84	At Astatine 85	Rn Radon 86
Fr Francium 87	Ra Radium 88	Ac Actinium 89 •																

H
Hydrogen 1

This line divides the metals from the non-metals.

x Lanthanide series

• Actinide series

Ce Cerium 58	Pr Praseodymium 59	Nd Neodymium 60	Pm Promethium 61	Sm Samarium 62	Eu Europium 63	Gd Gadolinium 64	Tb Terbium 65	Dy Dysprosium 66	Ho Holmium 67	Er Erbium 68	Tm Thulium 69	Yb Ytterbium 70	Lu Lutetium 71
Th Thorium 90	Pa Protactinium 91	U Uranium 92	Np Neptunium 93	Pu Plutonium 94	Am Americium 95	Cm Curium 96	Bk Berkelium 97	Cf Californium 98	Es Einsteinium 99	Fm Fermium 100	Md Mendelevium 101	No Nobelium 102	Lr Lawrencium 103

KEY

X
z

X = atomic symbol
z = atomic number

[1] Group I comprises the alkali metals.

[2] Group II comprises the alkaline–earth metals.

[3] Group VII comprises the halogens.

[4] Group 0 comprises the noble gases.

Know how to work out the atomic number of the first 20 elements using the periodic table.

PRACTICE

1 What is the middle block of the periodic table called?

2 What are the names of the metals in period 3?

THE BARE BONES

➤ Metals are mainly in groups I and II, and the central block of transition metals.

➤ Metals atoms are held together by their free-moving electrons.

➤ Alloys are usually a mixture of metals.

A Metallic bonding

KEY FACTS

1 In metals, the atoms are packed closely together, just like any solid.

2 The outer shell electrons of the atom are separated from the atoms in a sea of electrons.

3 The bond is the force between the positive metal ion and the electrons.

Remember
It is the forces between the positive metal ions and the free electrons in the metallic structure that cause the metallic bond to form.

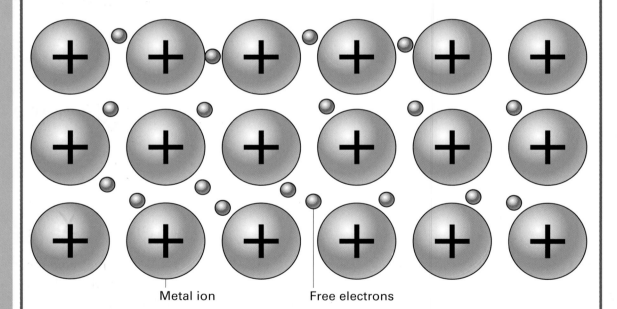

Metal ion   Free electrons

- From the diagram above you can see how the electrons are found in between the metal ions.

- The number of electrons in the outer shells of atoms of metals can vary. Copper is a particularly good conductor of electricity as it has a lot of electrons in its outer shell.

Q Which metal has the symbol Fe?

B   Properties of metals

1   Metals usually have <u>high melting</u> and <u>boiling points.</u>

2   They are <u>good conductors of electricity</u>.

- Electricity can pass through them easily as they have free electrons to carry the electricity.

3   They are <u>good conductors of heat.</u>

4   They are <u>ductile</u> and <u>malleable</u>.

- Ductile means they can be drawn into wires. Malleable means they can be shaped.

5 These properties are important reasons why metals are used for certain purposes. For instance, they are good conductors of heat and do not melt as they have high melting points. This makes them good for making saucepans and other cookware.

- As many metals are malleable and strong, they are used for many structures. For example, cars, planes, bridges.

- Many metals are useful as wires, for example: spokes of wheels, in fencing, and copper is used in electrical wiring.

Metals can
be shaped for
different uses.
What is the
name of this
property?

There are often exam questions which ask for the properties of metals.

C   Alloys

Alloys are mixtures of metals.
This usually makes the metal harder than their consituents.

Why is
carbon added to
iron?

- Alloys make use of the different properties of the different components. For instance brass uses the conductivity of copper and flexibility of zinc, to create wires, ornamental objects and cookware.

- Sometimes a metal is mixed with carbon. An important example is iron mixed with carbon to make steel.

1 What are the symbols for calcium, potassium and copper?

2 What properties do metals have which make them good for electrical wiring?

Group I – the alkali metals

THE BARE BONES

➤ Group I metals are the most reactive metals.
➤ These metals react with water and Group VII, the halogens.
➤ The reactivity of group I increases as you go down the group.

A Properties of the alkali metals

KEY FACT

1 The metals in group I are lithium (Li), sodium (Na) and potassium (K).

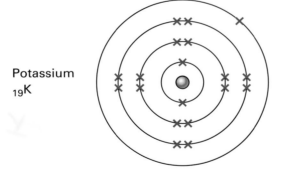

ALKALI METALS

Lithium
$_3$Li

Electronic arrangement

Sodium
$_{11}$Na

Potassium
$_{19}$K

Remember
They are called alkali metals because they produce an alkali (metal hydroxide) when they react with water.

The most common question on group I is about their reaction with water and what would be observed. Make sure you can list and explain the common reactions.

KEY FACT

2 Being in the same group they have similar properties.

Examples of common properties include:

- they are all shiny
- they can all be cut by a knife
- they must all be stored in oil – their single outer shell electron makes them so reactive they would otherwise react with the air
- they corrode easily because they are so reactive and react with the oxygen in air
- they are less dense than water, so when they react with water they float.

Q What ion does sodium form when it reacts?

B   Reactivity of the elements

1   When alkali metals react they <u>lose</u> the <u>one electron</u> in their outer shell.

2   Going down the group, the group I metals lose their electron more and more easily.

All alkali metals have just one electron in their outer shell. It is this that causes the similar properties within the group. Going down group I, each element has an extra shell, so that the outer electron is further away from the positive core. This means there is less force holding the outer electron, which makes it much easier to form a reaction with another element.

3   They form a positive ion when they react e.g. Li^+.

Which is more reactive: sodium (Na) or potassium (K)?

C   Reactions with water and the halogens

1   Group I metals violently react with water to produce hydrogen and a metal hydroxide.

Word equation:

metal + water → metal hydroxide + hydrogen

Example:

sodium + water → sodium hydroxide + hydrogen

2   Group I metals react with group VII elements (the halogens) to form metal halides.

Word equation:

group I metal + group VII halogen → metal halide

Example:

potassium + chlorine → potassium chloride

What compound is formed when potassium reacts with fluorine?

PRACTICE

1 Which is the least reactive group I metal?

2 What colour would universal indicator turn when group I metals react with water?

3 Why do group I metals have similar properties?

Transition metals

THE BARE BONES

➤ These metals are found in the central part of the periodic table.
➤ Their unique properties make them useful for many purposes.
➤ The transition metals form coloured compounds.

A Properties of transition metals

Remember
Transition metals are generally not very reactive metals.

Transition metals

The transition metals share a number of properties:

- They have high melting and boiling points.
- They are strong and hard.
- They have a high density.
- They are malleable and ductile.

Q What are the symbols for iron, zinc and copper?

Be prepared to compare the transition metals with Group I metals and their compounds.

These properties are similar to group I except transition metals have a higher density and their melting and boiling points are higher.

Transition metals are much stronger than most other metals.

B Uses of transition metals

KEY FACTS

1 These metals are often used as structural material, especially iron (Fe) and aluminium (Al).

- Copper is used for electrical wires.
- Zinc is used to galvanize iron.

2 Many of the transition metals are used as catalysts.

Q What is a catalyst?

- Iron is used as the catalyst in the Haber process to manufacture ammonia – an important industrial process. *For more on the Haber process see page 102.*

C Transition metal compounds

Most transition metal compounds are coloured.

It is this colouration on forming compounds that gives rise to the variety of gemstones and enables potters to create a vast array of glazes.

You will ave seen opper sulphate n science essons. What olour is it?

Examples of coloured compounds: copper chloride is blue/green, iron chloride is orange/brown, nickel chloride is green and cobalt chloride is blue.

- There are 2 compounds called copper oxide. Copper (I) oxide is red and copper (II) oxide is black.

D The rusting of iron

Corrosion is a chemical reaction, which occurs when a metal reacts with air and/or water. An example of this is iron rusting.

1   Iron is the only metal that <u>rusts</u>.

- Rust is the name of the compound formed when it corrodes.

2   Rust is <u>hydrated iron oxide</u>.

- Iron combines with oxygen in the air to form iron oxide, and then reacts with water to form hydrated iron oxide. Hydrated means water is present.

3   Painting, oiling and galvanising can prevent rusting.

- Painting and oiling prevent rusting because they are both waterproof, so the water cannot react with the iron.

- Galvanising works by coating the iron with a more reactive metal (usually zinc) so that only the zinc is in direct contact with the air and rain, and will react with them more easily than the iron.

    If the layer of zinc is scratched, the iron is exposed and will start to rust. This is called sacrificial protection – zinc is sacrificed to protect the iron.

What causes usting?

PRACTICE

1 Why is copper used for electrical wiring?

2 Which is more reactive potassium or copper?

3 What is the difference between group I compounds and transition metal compounds?

Metals and their reactivity

THE BARE BONES

➤ Reactivity of metals depends on their position in the reactivity series.

➤ A more reactive metal can displace a less reactive metal from its compound.

➤ Metals can react with oxygen, air, water and acids.

A The reactivity series

KEY FACT

Remember
Metals below hydrogen, e.g. copper, will not react with water or acid.

1 This is a list of metals in order of how reactive they are.

The reactivity series	
potassium K	most reactive
sodium Na	
calcium Ca	
magnesium Mg	
aluminium Al	
carbon C	
zinc Zn	
iron Fe	
tin Sn	
lead Pb	
hydrogen H	
copper Cu	
silver Ag	
gold Au	
platinum Pt	least reactive

Make up a mnemonic (rhyme) to remember the order of the reactivity series.

KEY FACTS

2 There are two non-metals in the list: carbon and hydrogen.

3 The reactivity series is determined by how vigorously the metal reacts with oxygen, water and acid.

Q What are the two products when potassium reacts with water?

4 Example Which metal will be the most reactive when it reacts with the air?

Potassium will be the most reactive as it is at the top of the reactivity series.

B Displacement reactions

1 In a metal compound, if a metal is lower in the reactivity series than another metal, displacement will take place.

Zinc will displace lead from lead oxide as zinc is more reactive than lead:

zinc + lead oxide → zinc oxide + lead

2 The higher the metal is in the reactivity series, the more difficult it is to extract.

- Elements above carbon are extracted using electrolysis.

- Elements below carbon can be extracted using reduction of the oxide.

  Example Write a word equation for the displacement reaction between magnesium and copper sulphate.

> As well as knowing the reactivity series, know what you will observe when a reaction takes place. A favourite is magnesium reacting with blue copper sulphate solution.

magnesium + copper sulphate → magnesium sulphate + copper

Will copper displace zinc from zinc sulphate?

C General reactions of metals

1 Metals react with oxygen in the air to produce a metal oxide

magnesium + oxygen → magnesium oxide

2 Metal react with water to produce metal hydroxide and hydrogen

calcium + water → calcium hydroxide + hydrogen

What gas is produced when magnesium reacts with sulphuric acid?

3 Metals react with acid to produce a salt and hydrogen

zinc + hydrochloric acid → zinc chloride + hydrogen

For more on acids see page 90

PRACTICE

1 Put these metals in order of reactivity with water:

Cu, Mg, Zn, Fe, Sn.

2 What would you observe if magnesium was added to blue copper sulphate?

3 If an iron nail is put into lead sulphate solution, what would be produced?

THE BARE BONES

➤ Group VII is made up of the halogens.
They are the reactive non–metals.
➤ Their reactivity decreases as you go down the group.
➤ Group VIII are called the noble or inert gases. They do not react.

A Group VII – the halogens and their properties

Remember
Noble gases are sometimes called the inert gases.

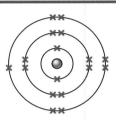

$_9$F
electron
arrangement

$_{17}$Cl
electron
arrangement

KEY FACTS

1 The halogens are coloured.

At room temperature, chlorine (Cl) is a **pale green gas**, bromine (Br) is a **red/brown liquid** and Iodine (I) is a **purple solid**.

2 They have <u>low melting and boiling points</u> compared to metals, and are poor conductors of heat and electricity.

The halogens are **diatomic molecules**. They are molecules of two atoms and are written as Cl_2, Br_2 and I_2.

Q What state would you predict fluorine will be at room temperature?

B Reactions of the halogens

KEY FACTS

1 The reactivity of the halogens decreases as you go down a group. A more reactive halogen will <u>displace</u> a less reactive halogen.

Chlorine will displace bromine from sodium bromide, as chlorine is more reactive than bromine.

2 When the halogens react they <u>gain 1 electron</u> and form a <u>negative ion</u>.

3 They react with hydrogen to form <u>hydrogen halides</u>, which form an <u>acid</u> in water.

Word equation hydrogen + chlorine → hydrogen chloride
hydrogen chloride + water → hydrochloric acid

Q How do you write the formula for a bromide ion?

C Electrolysis of sodium chloride solution

1 Sodium chloride (salt) is found in large quantities in the sea and in underground deposits.

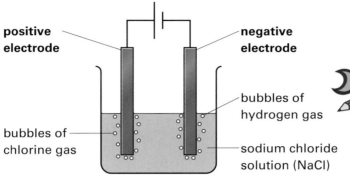

positive electrode

negative electrode

bubbles of hydrogen gas

bubbles of chlorine gas

sodium chloride solution (NaCl)

Know the uses of the different non-metal elements – an easy way to gain marks.

2 When sodium chloride solution is <u>electrolysed</u>, hydrogen is produced at the negative <u>electrode</u> and chlorine is produced at the positive <u>electrode</u>.

3 Chlorine is used for many purposes, such as:

- to **manufacture disinfectant** and **bleach**
- to **kill bacteria** in drinking water and in swimming pools.

The positive test for chlorine is if it **bleaches damp litmus paper**.

Does chlorine conduct electricity?

D Group VIII – the noble gases

The noble gases do not react, due to a full outer electron shell.

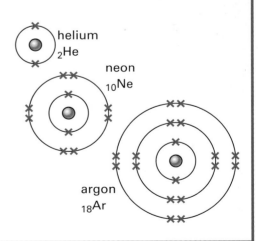

helium $_2$He

neon $_{10}$Ne

argon $_{18}$Ar

- They are mainly used to fill light bulbs and lamps because they are unreactive and so are not affected by the heat formed by the light.
- Helium is used for balloons, as it is less dense than air.
- Krypton is used for lasers.

What state is neon at room temperature?

PRACTICE

1 How many electrons do the halogens have in their outer shell?

2 Explain the following word equation:

chlorine + sodium bromide ➔ sodium chloride + bromine

3 Why are the noble gases unreactive?

Ionic bonding

➤ Bonding occurs when atoms of elements join together.

➤ When electrons are transferred between one atom and another, this is called <u>ionic bonding</u>.

A Why bonding occurs

KEY FACT

> Atoms bond together so that they have full outer shells of electrons.

Only electrons are involved in bonding. They can be shared or transferred from one atom to another.

Q Why are bonds formed in a compound?

B Ionic bonding

KEY FACT

1 Ionic bonding occurs between a <u>metal</u> and a <u>non-metal</u>.

2 Electrons are transferred from the metal to the non-metal.

3 The metal loses electrons to form positive ions and non-metals gain electrons to form negative ions.

4 You will need to show how atoms join together, using dot and cross diagrams, for the ions formed in sodium chloride, magnesium oxide and calcium chloride.

5 Example Here is a dot and cross diagram to show how magnesium oxide is formed:

Remember
The number of electrons lost or gained by the atoms is the number on the charge. Calcium loses two electrons and forms a Ca^{2+} ion.

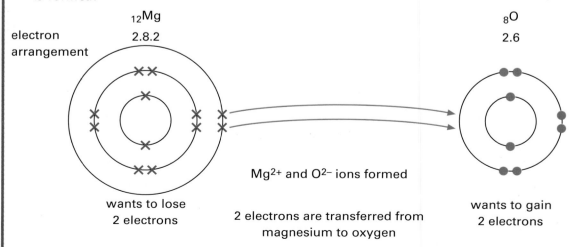

$_{12}Mg$
electron arrangement 2.8.2

$_8O$
2.6

Mg^{2+} and O^{2-} ions formed

wants to lose 2 electrons

2 electrons are transferred from magnesium to oxygen

wants to gain 2 electrons

Q Name an ionic compound.

c Properties of ionic compounds

1 The ionic bond is the attraction of the positive ions to negative ions; it forms a giant structure of ions. This is called a lattice structure.

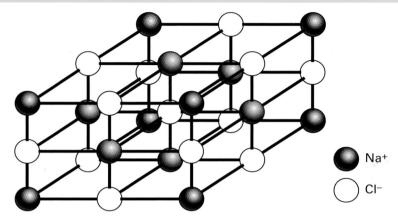

Na^+

Cl^-

As you can see from the diagram, each positive ion is surrounded by negative ions and each negative ion is surrounded by positive ions. The force between each ion is called an **electrostatic force**.

2 The attraction between the negative and positive ions is a strong force. Therefore, ionic compounds have high melting and boiling points.

3 Ionic compounds will conduct electricity when dissolved in water, or when melted, as the ions are free to move.

Solid ionic compounds will not conduct electricity because the ions are not free to move around.

Most ionic compounds are soluble in water and will therefore, once dissolved, conduct electricity.

Why doesn't n ionic solid onduct lectricity?

When asked to draw a dot and cross diagram to show bonding, ensure you have dots on one type of atom and crosses on the other type of atom.

1 What are the symbols for the sodium and chloride ions? What is the formula for sodium chloride?

2 Which type of particle is involved in ionic compounds?

3 Why do ionic compounds have high melting points?

Covalent bonding

THE BARE BONES

➤ Covalent bonding occurs between atoms of non-metals.
➤ Electrons are shared between one atom and another.
➤ Covalent compounds have low melting and boiling points, and they are poor conductors of electricity.

A Covalent bonding occurs between non-metals

KEY FACT

Covalently bonded substances are called <u>molecules</u>.

Q Does carbon dioxide involve ionic or covalent bonding?

Examples of covalently bonded substances are water, ammonia, hydrogen, hydrogen chloride and methane.

All these substances contain non-metals. Ammonia contains nitrogen and hydrogen, and methane contains carbon and hydrogen.

B How covalent bonds are formed

KEY FACT

1 Covalent bonds are formed when atoms share electrons.

Remember
You can also show a covalent bond by a line between atoms.

H–N–H
|
H

You will need to show how atoms join, using dot and cross diagrams like these:

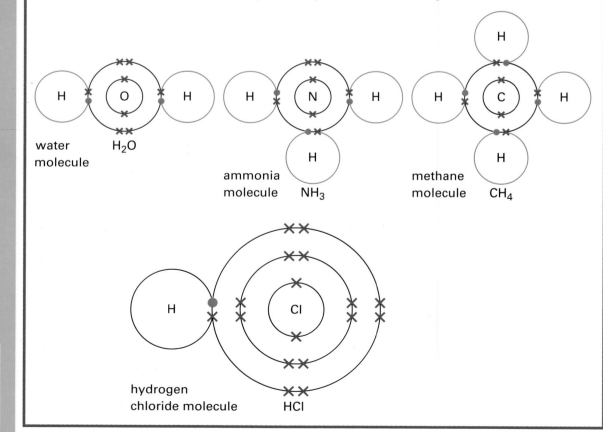

water molecule H_2O

ammonia molecule NH_3

methane molecule CH_4

hydrogen chloride molecule HCl

Q Do covalent substances contain ions?

C   Properties of covalent substances

1   The covalent bond, which holds the molecule together, is strong, but the force between the molecules is weak.

2   Covalent substances have low melting and boiling points and are usually gases and liquids at room temperature.

3   Covalent substances are poor conductors of electricity as covalent substances do not contain ions.

Q What would ou predict vould be the tate of methane at oom temperature?

When asked to draw a dot and cross diagram to show covalent bonding, ensure that you show that the electrons are being shared. The dots or crosses much be on both circles.

D   Giant covalent substances

1   The are two <u>giant covalent substances</u>, you need to know for the exam: <u>graphite and diamond</u>.

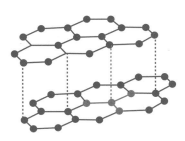   

Giant structure of graphite   Giant structure of diamond

• Diamond is an extremely hard substance.

• Graphite is unusual as it conducts electricity. This is because of the electrons, which are free to move in-between the layers.

Q Compare the nelting points f ammonia and liamond.

1   What is the difference between the formation of ionic and covalent bonds?

2   Draw a dot and cross diagram to show the formation of a chlorine molecule.

Chemical reactions

THE BARE BONES

➤ Decomposition reactions break down compounds.
➤ Combustion reactions occur when a substance reacts with oxygen.
➤ Oxidation and reduction reactions are opposite types of reaction.

A Decomposition reactions

KEY FACT

1 Decomposition is a reaction where a substance is decomposed or broken down.

2 There are three main types of decomposition reactions:

- **thermal decomposition** – where substances are broken down by heat
- **electrolytic decomposition** – where substances are broken down by electricity
- **catalytic decomposition** – where a catalyst breaks down substances.

3 Example
When calcium carbonate is heated it thermally decomposes.

calcium carbonate → calcium oxide + carbon dioxide

Q What type of decomposition happens if an electric current is passed through a substance?

Questions on different types of reaction could coincide with questions on electrolysis, extraction of metals and fuels.

B Combustion reactions

KEY FACTS

1 Combustion reactions burn substances in oxygen.

2 Combustion is an important reaction for fuels. When they burn they produce carbon dioxide and water.

These reactions are usually exothermic. This means they produce heat. Fuels need to produce heat when they are combusted.

3 Example
Methane burns in oxygen to produce carbon dioxide and water.

methane + oxygen → carbon dioxide + water

Q What gas are substances burnt in during a combustion reaction?

c Oxidation and reduction

emember

1L RIG

Oxidation
s
Loss of electrons
eduction
s
Gain of electrons

1 Oxidation has two definitions.

- *Oxidation is a reaction where a substance combines with oxygen.*
- *Oxidation is a reaction where electrons are removed from the element in a reaction.*

2 Reduction has two definitions.

- *Reduction is a reaction where oxygen is removed from a substance.*
- *Reduction is a reaction where electrons are added to the element in a reaction.*

Notice the definitions are the opposite of each other.

3 Example In the following reaction, is the iron oxidised or reduced?

iron oxide + carbon → iron + carbon dioxide

The iron is reduced, as oxygen is removed from the iron oxide.

In many reactions, both oxidation and reduction occurs. These are called redox reactions.

4 Example

lead oxide + carbon → lead + carbon dioxide

The lead oxide is reduced to lead as oxygen is removed.

The carbon is oxidised to carbon dioxide, as the oxygen is added to the carbon.

So both oxidation and reduction take place, so this is a redox reaction.

What
appens during
n oxidation
eaction?

PRACTICE

1 If a sodium atom loses an electron to form a sodium ion, is this reduction or oxidation?

2 What are the two products formed when a fuel is combusted in oxygen?

3 Write a word equation for the thermal decomposition of copper carbonate.

Writing chemical equations

THE BARE BONES
- ➤ Word equations contain the names of the reactants and products.
- ➤ Symbol equations need a balanced number of atoms on both sides.
- ➤ Each substance in a symbol equation needs a state symbol.

A Writing word equations

You'll often be asked to write a word equation to represent a chemical reaction. It is also useful to write down the word equation first when asked for a symbol equation.

Example

Potassium reacts with water to produce potassium hydroxide and hydrogen.

1 Write a word equation for this reaction:

potassium + water → potassium hydroxide + hydrogen

2 Write the chemical reaction using the word equation above:

$2K + 2H_2O \rightarrow 2KOH + H_2$

> **Q** Convert to a word equation:
>
> $2Mg + O_2 \rightarrow 2MgO$

> If you are asked for a symbol equation, often it would be worth more than one mark. If you cannot do all of it, you may gain a mark for the part you can do.

B Writing formulae for compounds

KEY FACTS

1 Each element has a combining power, which tells you the number of electrons the element needs to lose, gain or share when it forms a compound.

When elements combine, the combining powers of both elements in the compounds must be equal.

2 The combining power depends on which group it is in the periodic table.

- Group I to IV = group number.
- Group V to VIII = 8 – group number.

Example To work out the formula of magnesium chloride:

1 Magnesium is in group II, so it has a combining power of 2.

2 Chlorine is in group VII, so it has a combining power of 8–7 = 1.

3 To make the combining power equal you will need two chlorine atoms.

4 The number of each type of atom is put after the symbol. So one atom of Mg combines with 2 atoms of Cl, giving the formula $MgCl_2$.

> **Q** What is the formula for sodium chloride?

C Balancing equations

Remember
When you are balancing an equation you can only put numbers in front of symbols. If you put numbers after the symbols, you will be altering the compound.

Step 1:
Write out the word equation.

sodium + chlorine = sodium chloride

Step 2:
Work out the formula. The combining powers are equal, so one atom of sodium combines with one atom of diatomic chlorine.

$Na + Cl_2 = NaCl$

formula for chlorine gas   combining power = 1   combining power = (8 − 7) = 1

Step 3:
Balance the equation. To balance the equation you need to make sure there are the same number of atoms of each elements on both sides of the equals sign. When you are balancing, the number goes in front of the symbols.

Na +	Cl_2	= $NaCl$	not balanced
1'Na'	2'Cl'	1'Na', 1'Cl'	need 2 'Cl' on RHS
Na +	Cl_2	= $2NaCl$	not balanced
1'Na'	2'Cl'	2'Na', 2'Cl'	need 2 'Na' on LHS
$2Na$ +	Cl_2	= $2NaCl$	balanced
2'Na'	2'Cl'	2'Na', 2'Cl'	

Step 4:
Put in the state symbols
The state symbols show whether the substance is solid, liquid or gas (what 'state' the substance is in). You should know the following state symbols:
(l) = liquid; (s) = solid;
(g) = gas; (aq) = in solution.

$2Na(s) + Cl_2(g) = 2NaCl(s)$

Check carefully in exam questions to see whether a word or symbol equation is asked for.

Q What is the state symbol for a solution?

PRACTICE

1 What is the formula for carbon dioxide?

2 Balance the following equation:

$H_2 + O_2 \rightarrow 2H_2O$

3 Put the state symbols in the following equation:

$CaCO_3 \rightarrow CaO + CO_2$

Chemical calculations

THE BARE BONES

➤ You can calculate the relative formula mass of a compound by adding up the individual atomic masses of the elements.

➤ From a compound's formula it is possible to calculate the percentage of the elements.

A   Relative atomic mass

KEY FACTS

1   As one atom weighs almost nothing, the scale for measuring atoms is <u>relative to other atoms</u> and so is called the <u>relative atomic mass</u>.

2   The relative atomic mass of atoms of elements (A_r) is <u>based on a scale</u> where <u>carbon</u> has a mass of 12 units.

3   An <u>ion has the same relative atomic mass as the atom</u> from which it originated.

Q What is the relative atomic mass of a chloride ion?

Examples The relative atomic masses of the following ions are:
a) the sodium ion Na^+ = 23   b) the oxide ion O^{2-} = 16.

• The charge on the ion makes no difference to the relative atomic mass of the atom.

B   Calculating relative formula mass

KEY FACTS

1   The relative formula mass is the <u>mass of a substance</u> <u>which is not a single element</u>.

2   To calculate the relative formula mass you <u>add up the individual relative atomic masses</u> of the elements in the substance.

Calculating the relative formula mass is straightforward, but remember, the number after the symbol tells you how many of that atom there are in the formula.

3 Example What is the relative formula mass of sodium chloride NaCl?

• Relative atomic masses (A_r) of components of sodium chloride NaCl:

Sodium A_r = 23
Chlorine A_r = 35.5

• Relative formula mass of NaCl = 23 + 35.5   = 58.5

Remember
You do not have to know the relative atomic masses; they will be given to you in the exam paper.

B

Example What is the relative formula mass of water H_2O?

- Relative atomic masses (A_r) of components of water H_2O:

  Hydrogen $A_r = 1$
  Oxygen $A_r = 16$

- Relative formula mass of $H_2O = (2 \times 1) + 16$
  $$= 18$$

Q What is the relative formula mass of oxygen molecule O_2 (A_r O = 16)?

C *Calculating percentages of elements in a compound*

Example What is the percentage of carbon in carbon dioxide (CO_2)?

KEY FACT

1 To calculate the percentage of any element in a compound, you first need to calculate the relative formula mass of the compound.

- Relative atomic masses (A_r) of components of CO_2:

  Carbon $A_r = 12$
  Oxygen $A_r = 16$

- Relative formula mass of $CO_2 = 12 + (2 \times 16)$
  $$= 44$$

KEY FACT

2 Then you can work out the percentage of each element.

- % of element = $\dfrac{\text{mass of the element}}{\text{relative formula mass}} \times 100$

- % of carbon = $\dfrac{12}{44} \times 100$
  $$= 27.3\ \%$$

Q What is the percentage of calcium in calcium carbonate $CaCO_3$?

Usually, calculations questions are worth two or three marks, so even if you can't complete all the calculations, do as much as you can and gain some of the marks.

PRACTICE

1 What is the mass of a magnesium ion Mg^{2+} (A_r Mg = 24)?

2 What is the relative formula mass of methane CH_4 (A_r C = 12 H = 1)?

3 What is the percentage of hydrogen in ammonia NH_3 (A_r N = 14 H = 1)?

THE BARE BONES

➤ Acids have a pH of less than 7.
➤ Acids react with metals to produce a salt and hydrogen.
➤ Acids react with a metal carbonate to produce a salt, carbon dioxide and water.

A Acids and the pH scale

KEY FACT

1 The <u>pH scale</u> is used to predict whether a substance is <u>acid</u>, <u>alkali</u> or <u>neutral</u>.

Remember
Universal indicator is used to measure pH.

pH scale

1	2	3	4	5	6	7	8	9	10	11	12	13	14

strong acid   weak acid   weak alkali   strong alkali

neutral

red   orange   yellow   green   blue   purple

KEY FACT

2 Acids have a pH of less than 7.

The **lower the number**, the **stronger the acid**. A weak acid will have a pH of just below 7, while a strong acid will have a pH of between 1 and 3.

KEY FACT

3 <u>All acids contain a positive hydrogen ion H^+.</u> The common laboratory acids are hydrochloric acid (HCl), sulphuric acid (H_2SO_4) and nitric acid (HNO_3).

Hydrochloric acid and nitric acid contain one hydrogen ion, whereas sulphuric acid contains two hydrogen ions.

KEY FACT

4 The common household acids are citric acid, which is found in oranges and lemons, and vinegar.

- The household acids (citric acid and vinegar) are weak acids. They have a pH of 4–6.
- Hydrochloric acid, which is in your stomach, is a strong acid and has a pH of 1.

Q What is the pH of a strong acid?

Naming salts produced in a reaction with an acid, or filling in gaps in word equations of reactions with acids, are regular questions on the exam papers.

B   Reactions of acids with metals

1   Most acids react with metals to produce a salt and hydrogen.

The general word equation is: metal + acid → salt + hydrogen

Example

zinc + hydrochloric acid → zinc chloride + hydrogen

2   To test if hydrogen is produced – a lighted splint 'pops'.

3 To name the salt formed:
- the first part is the same as the metal
- the second part depends on the acid used.

Examples
- Hydrochloric acid produces chlorides.
- Sulphuric acid produces sulphates.
- Nitric acid produces nitrates.

Name the salt produced when calcium carbonate reacts with sulphuric acid.

C   Reaction of acids with carbonates

1   Acids react with metal carbonates to produce a salt, carbon dioxide and water.

The general word equation is:

acid + metal carbonate → salt + carbon dioxide + water

Example

copper + nitric acid → copper nitrate + carbon dioxide + water

2   You can test the carbon dioxide by observing if limewater turns cloudy.

What gas is produced when magnesium reacts with any acid?

PRACTICE

1 What ion is contained in all acids?

2 What are the three products formed when copper carbonate reacts with hydrochloric acid?

3 Fill in the gap in the following word equation:

magnesium + _____ acid → magnesium nitrate + _____.

Bases and neutralisation

THE BARE BONES

➤ Bases are metal oxides and hydroxides, with a pH greater than 7.
➤ Neutralisation is when an acid reacts with bases to produce salt and water.
➤ Neutralisation is useful in cleaning teeth and curing indigestion.

A Bases and alkalis

KEY FACTS

1 Bases are metal oxides and hydroxides.

An example of a base is copper oxide (CuO) or sodium hydroxide (NaOH).

2 A soluble base is called an <u>alkali</u>. A common laboratory alkali is sodium hydroxide.

• This means alkalis are bases, which dissolve in water.

3 Alkalis contain hydroxide ions OH^-.

4 Most common household alkalis are used for cleaning.
Examples are: washing powders, soap and oven cleaner.
These are strong alkalis with a pH of 11.

Q What is the difference between an alkali and a base?

You can guarantee a question on the exam paper concerned with neutralisation. Learn the general equation and how it is used in everyday life.

B Neutralisation reaction

KEY FACTS

1 When an acid reacts with a base, <u>neutralisation occurs</u>.

• Hydrochloric acid works in your stomach to digest food, but too much can cause indigestion. Medicines to help indigestion contain an alkali or a base. Common products often contain aluminium hydroxide.

2 In a neutralisation reaction, a <u>salt</u> and <u>water</u> are formed.

The general word equation for this is:

acid + base → salt + water

3 Examples

sulphuric acid + copper oxide → copper sulphate + water

hydrochloric acid + sodium hydroxide → sodium chloride + water

• The hydrogen ion (H^+) in the acid combines with the hydroxide ion (OH^-) in the alkali to produce water:

$H^+ + OH^- = H_2O$

Q Which of the following is a base: sodium chloride, potassium oxide, magnesium nitrate?

C Application of neutralisation

Neutralisation reactions occur in everyday life.

Examples:

1 You can take an alkali to neutralise the excess acid which causes indigestion.

2 When you clean your teeth, the acid on your teeth is neutralised by brushing them with toothpaste that contains an alkali.

3 If soil is too acidic an alkali, such as lime (calcium oxide), would neutralise it.

4 Stinging nettles contain an acid; use the alkali in a dock leaf, which can be found nearby, to neutralise the acid in the sting.

5 Insect bites can be cured using neutralisation reactions. A bee sting is acid and is neutralised by a base, such as baking soda (sodium hydrogen carbonate). A wasp sting is an alkali and is neutralised by an acid, such as lemon juice or vinegar.

What liquid
always
roduced in a
eutralisation
action?

RACTICE

1 What ion is contained in an alkali?

2 What type of substance do you take to cure indigestion?

3 Complete the following word equation

sodium hydroxide + hydrochloric acid →

Rates of reaction I

THE BARE BONES

➤ Rate of reaction is a measure of change in a reaction over time.
➤ There are various factors that affect the rate of reaction.
➤ Factors that affect rate can be explained using the collision theory.

A Measuring rate

KEY FACTS

1 There are two main ways of measuring rates.

2 You can measure the amount of product formed.

3 You can measure the decrease in mass of the reactants.

Example What do the following graphs show?

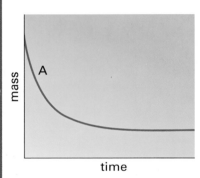

Graph A represents the amount of reactants changing with time and B represents the amount of product changing with time.

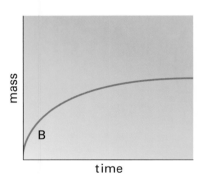

The reaction has stopped because there are no more reactants.

Q. What will happen to the mass if a gas is given off when the reaction takes place?

B Factors that affect rates

KEY FACTS

1 Increasing the temperature of the reactant increases the rate of the reaction.

2 Increasing the concentration of the reactant increases the rate of the reaction.

3 Increasing the surface area of the reactant increases the rate of reaction.

4 A catalyst increases the rate of a reaction without taking part in the reaction itself.

Q. What happens to rate graphs if the reaction goes faster?

c   The collision theory

Y FACT

1 The collision theory states that particles will react if they collide with sufficient energy.

If you increase the temperature of the reactants they will have more energy and will have more chance of a successful collision.

Y FACT

2 When you increase the concentration of a reactant, there are more particles which can collide successfully.

• If the reactants are in a solution and the solution is concentrated, the amount of water is small, so the reactant particles are more likely to collide with each other rather than the water.

• If the reactants are in a solution and the solution is dilute then there are more water particles so there is less chance of the reactant particles colliding, as there are many more particles of water.

Y FACT

3 A catalyst lowers the amount of energy needed for a successful collision. The energy needed for a successful collision is called the <u>activation energy</u>.

Example In a reaction between magnesium and an acid, it will react more slowly if water has been added to the acid. This is because the acid is less concentrated, which means there is less chance of a collision of an acid with magnesium.

4 If you increase the surface area, there is more chance of successful collisions. With a large surface area, there are more particles exposed, so there is a greater chance of a collision of reacting particles occurring. If the surface area is small, then many reacting particles are not exposed so they cannot collide.

Examples The smaller you cut up potatoes, the faster they will cook.

In a reaction with an acid, a lump of calcium carbonate would react more slowly than powdered calcium carbonate because is has a smaller surface area.

Define a
talyst.

Make sure you are able to link the factors that affect reaction rates to the collision theory.

RACTICE

1 Why does a catalyst speed up a reaction even at low temperature?

2 Fill in the gaps:
A reaction goes_____ when the concentration of a _____is increased. It also goes _____ when the _____ is raised.

Rates of reaction II

THE BARE BONES

➤ Graphs are used to show the rates of reaction.
➤ Reactions can be exothermic or endothermic.
➤ Reactions that go both ways are called reversible reactions.

A Drawing and interpreting graphs involving rates

KEY FACTS

1 Time is always represented on the x-axis (horizontal axis).

2 Reaction rate graphs are drawn as line graphs.

3 Lines of best fit (not joining all the points) are necessary to interpret the graphs.

Q In a graph showing rate what should be on the x-axis?

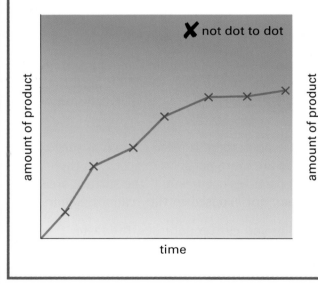

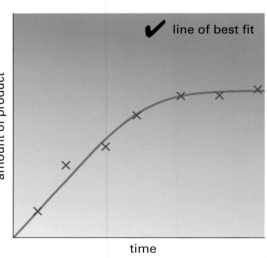

B Exothermic and endothermic reactions

KEY FACTS

1 A reaction which gives out heat to the surroundings is an <u>exothermic</u> reaction.

2 A reaction which takes heat in from the surroundings is called an <u>endothermic reaction</u>.

In any reaction, first **bonds are broken** (this is an endothermic process), then **new bonds are made** (this is an exothermic process).

B

emember

he test for
ater: blue
obalt chloride
aper turns
ink.

**Is bond-
naking
xothermic or
ndothermic?**

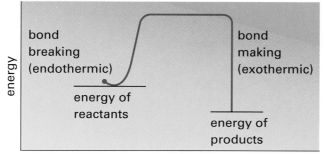

Example During a reaction between methane and oxygen the temperature of the surroundings will increase. This is because it is an exothermic reaction.

When drawing graphs, check the points you have plotted carefully. They will usually have put the scale in for you.

C Reversible reactions

'Y FACT

1 A reversible reaction is a reaction that can go both ways. The reactants are converted to products and the products can be converted back to the original reactants.

For example, the thermal decomposition of hydrated copper sulphate is a reversible reaction (hydrated means 'containing water', anhydrous means 'without water'):

hydrated copper sulphate $\rightleftharpoons$ anhydrous copper sulphate + water

$$CuSO_4.5H_2O \rightleftharpoons CuSO_4 + 5H_2O$$

blue crystals   white powder

**What does
$\rightleftharpoons$ mean?**

To turn white copper sulphate blue again, just add water.

PRACTICE

1 A solution of hydrogen peroxide (H_2O_2) decomposes slowly to water and oxygen.

a) Complete the following word equation:

hydrogen peroxide $\rightarrow$ _____ + _____.

b) This reaction is speeded up by adding manganese (IV) oxide. What name is given to a substance that speeds up a reaction without being used up?

c) Tina found the mass of hydrogen peroxide at various times and worked out the loss in mass. Her results are shown in the table. Using graph paper, plot the points and draw a line graph of loss in mass against time.

time (minutes)	mass (g)	loss in mass (g)
0	40.1	0.0
2	39.5	0.6
4	39.3	0.8
6	39.2	0.9
8	39.1	1.0
10	39.1	1.0
12	39.1	1.0

d) Explain in one sentence why the reaction slowed down.

Enzymes

THE BARE BONES
- ➤ Enzymes are biological catalysts – they speed up reactions.
- ➤ Enzymes are used in fermentation.
- ➤ Enzymes are used in detergents and yoghurt-making.
- ➤ Enzymes are used in baby foods and in slimming foods.

A What are enzymes?

KEY FACTS

1 Enzymes are proteins found in all living things.

2 Enzymes speed up the rate of reactions.

Enzyme activity can be affected by temperature and chemical conditions.

Example Salivary amylase, which is found in saliva, works at 37°C and in slightly alkaline conditions (pH = 7.5).

Q Name the enzyme found in saliva.

There is likely to be a question on use of enzymes. This is a new part of the chemistry course.

B Fermentation

KEY FACTS

1 The process of fermentation involves yeast, which is a living organism and contains an enzyme.

- Fermentation is used in beer- and bread-making.

2 In beer-making, sugar and yeast are converted to alcohol and carbon dioxide.

This can be represented as a word equation:

$$\text{sugar} + \text{yeast} \rightarrow \text{alcohol} + \text{carbon dioxide}$$

- This is also the process used to make wine. You can tell the fermentation reaction is taking place because the wine will bubble as the carbon dioxide is given off.
- The test for carbon dioxide is to use limewater. If the limewater turns cloudy, this is a positive result for carbon dioxide.
- The process of distillation is used to separate the alcohol when fermentation is complete.

Q Fermentation produces carbon dioxide. How could you test for this gas?

KEY FACT

3 In bread-making, yeast produces carbon dioxide, which causes the bread to rise.

- The yeast converts the starch in the flour into sugar and then fermentation occurs.

C Use of enzymes in the home and in yoghurt-making

KEY FACT

1 Biological enzymes are used in washing powder, which will digest the dirt on the clothes. They are proteases and lipases.

• The proteases will digest proteins and the lipases will digest fats.

KEY FACT

2 Yoghurt is made from milk and a living bacterium, which contains an enzyme.

• The enzyme in the bacterium converts the sugar in the milk to lactose.

• When making yoghurt the milk is pasteurised first. This involves heating the milk so that any bacteria are killed. Then different bacteria is added, containing an enzyme, which converts the sugar into a substance that makes the milk thicken.

Enzymes are proteins, which are highly sensitive to factors that affect their structure, such as temperature and pH. This is because their molecular structure is vital to their function.

Example If a biological washing powder was used in a cycle where the temperature of the water was too high, the enzymes would break down and not clean the clothes properly.

Q What does the lipase in biological detergent digest?

D Use of enzymes in industry

KEY FACT

1 Enzymes called proteases are used in baby foods to make the proteins easier to digest.

• The enzyme in the baby food has already broken down some of the protein so it doesn't need to be digested any more.

KEY FACTS

2 The enzyme isomerase is used to convert glucose to fructose. Fructose is much sweeter than glucose and can be used in smaller amounts in slimming foods.

3 Enzymes are often used in the industrial processes to cut down on cost, as many enzymes allow reactions to take place at lower temperatures.

Q What does 'pasteurised' mean?

PRACTICE

1 What is present in yeast that causes the fermentation of sugar?

2 What is the name of the enzyme used in detergent to digest proteins?

3 How can enzymes cut down costs in industry?

Useful products from oil

THE BARE BONES

➤ Crude oil is a mixture of many compounds, called hydrocarbons.
➤ Crude oil can be separated by fractional distillation.
➤ Combustion of hydrocarbons produces carbon dioxide and water.
➤ Large hydrocarbons can be cracked to form smaller hydrocarbons.

A What is crude oil?

KEY FACTS

1   Crude oil is formed from dead sea creatures, which have been trapped between sediments over millions of years.

2   Crude oil is a <u>fossil fuel</u>.

3   Crude oil contains a large number of compounds, called <u>hydrocarbons</u>.

4   Hydrocarbons are compounds containing <u>carbon</u> and <u>hydrogen</u> only.

Q Is oil a fossil fuel?

Example Ethanol has a formula C_2H_5OH. Is it a hydrocarbon?
Answer No, because it contains oxygen.

B How is crude oil separated?

KEY FACT

1   Crude oil can be separated using fractional distillation as the separate hydrocarbons have different boiling points.

Remember
To remember fossil fuels, use 'COG' (coal/oil/gas).

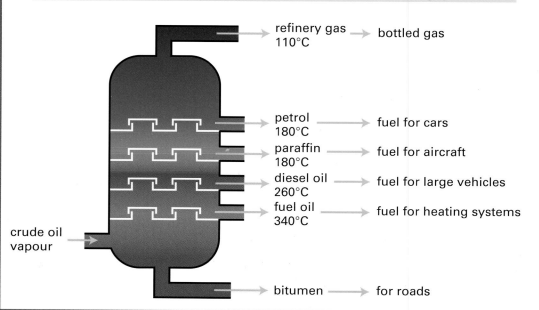

B

2 The hydrocarbons with the highest boiling points are distilled at the bottom of the fractionating column. Going up the column, the boiling points decrease.

sets alight easily
flows easily

more volatile
changes into
gas easily

largest molecule
(more carbon
atoms in it)

highest boiling point

The compounds formed at the bottom of the fractionating column are more viscous than those formed at the top. Viscous means thicker and flows less easily.

The products at the top of the column are gases and the products at the bottom of the column are solids.

Cracking can be called thermal decomposition, explain why.

Often in questions about hydrocarbons you will be asked to balance a combustion equation.

C *Hydrocarbons*

1 Hydrocarbons combust in oxygen to produce carbon dioxide and water.

hydrocarbon + oxygen → carbon dioxide + water

methane + oxygen → carbon dioxide + water

2 Large hydrocarbon molecules can be cracked to produce smaller, more useful molecules, using catalysts.

Products of cracking can be used as fuel and to make plastic.

3 Two useful plastics are polyethene (used in bags and bottles) and polypropene (used in crates and ropes).

In what part of the fractionating column are compounds with low boiling point likely to be formed?

PRACTICE

1 Which two elements are found in compounds in crude oil?

2 Methane is a hydrocarbon with a formula CH_4. Write a balanced equation for the combustion of methane.

Ammonia and fertilisers

THE BARE BONES

➤ Ammonia is manufactured from nitrogen and hydrogen.
➤ Fertilisers are manufactured from ammonia.
➤ Fertilisers are useful in plant growth, but can be dangerous if used in excess.

A   Production of ammonia – the Haber process

KEY FACTS

1   Ammonia is manufactured from nitrogen and hydrogen. This is a reversible reaction, called the Haber process.

nitrogen   +   hydrogen   ⇌   ammonia
N_2 (g)   + $3H_2$ (g)   ⇌ $2NH_3$ (g)

2   The raw materials are air, methane and hydrogen. Nitrogen is obtained from the air and hydrogen is obtained from the reaction of steam with methane.

3   The gases are passed over an iron catalyst at 450°C and at a pressure of 200 atmospheres.

4   The ammonia is cooled and liquefied.

5   As it is a reversible reaction, some ammonia breaks down to nitrogen and hydrogen, which can be recycled and reacted again.

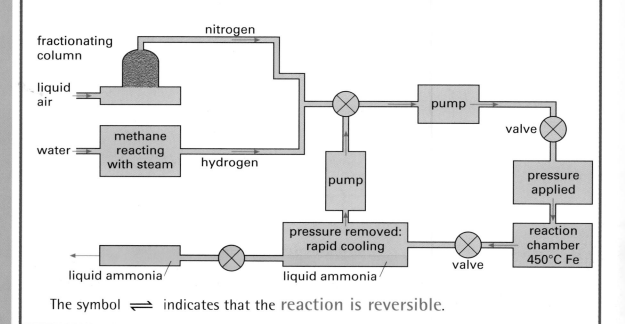

The symbol ⇌ indicates that the reaction is reversible.

Q What is the most important element in ammonium nitrate, when it is used as a fertiliser?

B Manufacture of fertilisers

1 Fertilisers contain <u>nitrogen</u> and improve the growth of crops. A good fertiliser should be <u>soluble in water</u>.

2 Ammonium nitrate (NH_4NO_3) is an important fertiliser, manufactured by reacting ammonia with nitric acid.

3 Nitric acid is made by reacting ammonia with oxygen to produce nitrogen monoxide. This is reacted with more water and oxygen to make nitric acid.

ammonia + oxygen → nitrogen monoxide

nitrogen monoxide + oxygen + water $\xrightarrow{\text{platinum catalyst}}$ nitric acid

4 Nitric acid is then reacted with ammonia to produce ammonium nitrate. The manufacture of ammonium nitrate from nitric acid and ammonia is a neutralisation reaction (an acid reacting with a base), as ammonia is a base.

Remember
he raw
aterials and
he conditions
or the Haber
rocess

Why is
ertiliser added
o soil?

C Too much fertiliser!

Too much fertiliser can cause contamination of drinking water.

plants   fertilisers in soil   animals
↓   ↓   ↓
———dissolves———
washed into rivers   washed into drains
↓   ↓
taken into water   taken into
containing plants (algae) drinking supplies
↓   ↓
rapid growth   too much nitrate in drinking water
↓   ↓
too many algae   converted to nitrites
↓   ↓
loads of bacteria grow   causes blood
on decaying algae   not to be able
↓   to carry oxygen
less oxygen in water   ↓
↓   called 'blue
animals in water die   baby' syndrome
↓
eutrophication

Fertiliser questions will often be linked with calculating the percentage of element in a fertiliser.

What causes
abies to have
lue baby
yndrome'?

1 How many different elements are there in ammonium nitrate NH_4NO_3?

2 a) How can the nitrogenous fertiliser put onto fields get into river water?
  b) What property of ammonium nitrate enables this to happen?
  c) What effect might the nitrogenous fertiliser have on the algae in the river?

Extraction of metals

THE BARE BONES
➤ Reactive metals can be extracted using electrolysis.
➤ Iron is extracted using a blast furnace.
➤ Copper is purified using electrolysis.

A Extraction of aluminium

KEY FACT

1 Metals that are <u>high in the reactivity series</u> are extracted using electrolysis.

Aluminium is extracted by electrolysis from its ore: aluminium oxide. Aluminium oxide is also known as bauxite and has to be molten so that the ore ions are free to move.

2 Cryolite is added to lower the temperature.

3 The electrodes are made of carbon.

4 Aluminium is formed at the negative electrode.

5 Oxygen forms at the positive electrode and reacts with the carbon electrode to form carbon dioxide. This is why the electrodes have to be replaced regularly.

carbon anode (+)
— bauxite in molton cryolite
— carbon lining cathode (–)
— molten aluminium

Q What electrode are the aluminium ions attracted to?

B Extraction of iron

KEY FACTS

1 In the reactivity series, metals that are less reactive than carbon can be extracted by reduction using carbon.

2 Iron (lower than carbon in the reactivity series) is extracted in a blast furnace.

• Iron ore is reduced by carbon dioxide.
• The raw materials are limestone (calcium carbonate), coke and iron ore.
• Limestone (calcium carbonate) is added to remove impurities and forms slag.

The reaction between coke (carbon) and oxygen is mainly responsible for producing the necessary high temperatures.

Remember

Negative ions are attracted to the positive electrode

Positive ions are attracted to the negative electrode.

Q Why is limestone added to the blast furnace?

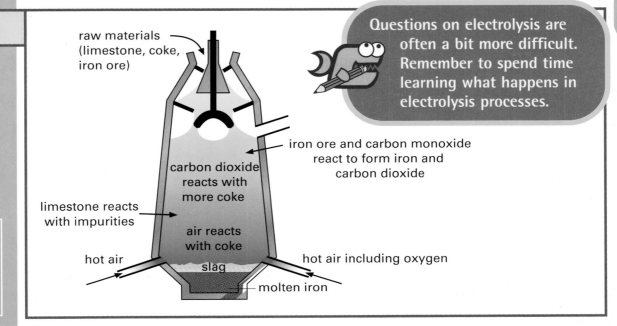

raw materials (limestone, coke, iron ore)

iron ore and carbon monoxide react to form iron and carbon dioxide

carbon dioxide reacts with more coke

limestone reacts with impurities

air reacts with coke

hot air

slag

hot air including oxygen

molten iron

Questions on electrolysis are often a bit more difficult. Remember to spend time learning what happens in electrolysis processes.

C Purification of copper

KEY FACTS

1   Some metals, such as copper, are purified by electrolysis.

2   The <u>positive</u> electrode is <u>impure copper</u> and the <u>negative</u> electrode is <u>pure copper</u>. <u>Copper ions</u> are in the solution.

Copper ions from the impure copper electrode **move** to the pure copper electrode. The **impurities** fall to the bottom of the container.

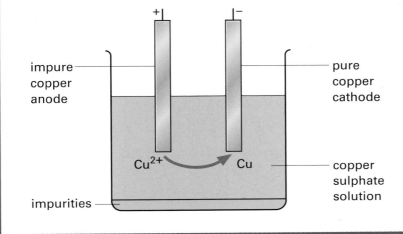

impure copper anode

pure copper cathode

Cu^{2+} Cu

copper sulphate solution

impurities

Q What electrodes are the copper ions attracted to?

PRACTICE

1   What method would you use to extract potassium from its ore?

2   What substance is used to reduce iron ore in the blast furnace?

3   What solution is used in the purification of copper?

Rocks and the rock cycle

THE BARE BONES
➤ Rocks can be: igneous, sedimentary or metamorphic.
➤ Rocks can be: weathered, eroded and transported.
➤ All these rocks and processes are connected through the rock cycle.

A Igneous rock

KEY FACTS

1 Igneous rocks are formed by <u>magma</u> (molten rock), which then cools to form a solid.

• The slower the rock cools, the larger the crystals formed are in the rock.

2 Igneous rocks formed outside the Earth's crust are called <u>extrusive igneous rocks</u>.

• An example of an extrusive igneous rock is basalt. It has small crystals because it cools quickly.

3 Igneous rocks formed inside the Earth's crust are called <u>intrusive igneous rocks</u>.

• An example of an intrusive igneous rock is granite. It has larger crystals because it cools slowly.

Q What are intrusive igneous rocks?

> You will often be asked to fill in gaps in diagrams on the different processes and types of rocks.

B Sedimentary rocks

KEY FACT

1 Sedimentary rocks are formed by <u>rock fragments</u> being <u>deposited</u> (sediments) and <u>compressed</u> together.

• Mudstone, limestone and sandstone are examples of sedimentary rock.

• Younger sedimentary rocks are formed on the top of older rocks.

KEY FACT

2 Fossils are often found in sedimentary rocks.

• Fossils are formed in the layers of rock beacuse as the plant or anuimal dies it is covered with layers of sediment, which eventually forms sedimentary rock.

KEY FACTS

3 The processes involved in forming sedimentary rocks are: <u>weathering</u>, <u>erosion</u>, <u>transportation</u>, <u>deposition</u> and <u>burial</u>.

Q What does erosion mean?

4 There are three types of weathering: <u>physical</u>, <u>chemical</u> and <u>biological</u> – the same as the three areas of science.

C Metamorphic rocks

1 Metamorphic rocks are formed when rocks are changed by heat and pressure.

2 In these circumstances, slate is formed from mudstone and marble is formed from limestone. These are examples of sedimentary rocks forming metamorphic rocks.

3 The word 'metamorphic' means 'transformed' or 'changed'. Many mountain ranges are made from metamorphic rocks, often formed because of the movement of the Earth's plates. This means the rocks are heated and put under pressure.

4 Metamorphic rocks are used for flooring (slate) and parts of buildings or statues (marble).

What does mestone form fter the effect f heat and emperature?

D The rock cycle

The cycle connects all the three rock types and processes.

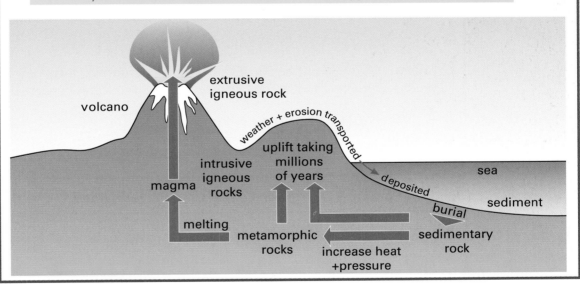

Which is the ldest layer?

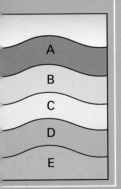

1 The diagram (right) gives an outline of how rocks are formed.
a) Name the rocks formed at A and B.
b) How is the sediment changed into sedimentary rock?
c) Suggest how the mountain may have been formed.

2 Which type of rock may contain fossils?

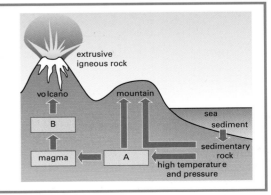

Changes in the atmosphere

THE BARE BONES

➤ Air is mainly nitrogen and oxygen.
➤ Over the past 200 million years the atmosphere has changed little.
➤ Changes in the atmosphere are partly due to the evolution of plants.
➤ The carbon cycle maintains the basic balance of the atmosphere.

A The atmosphere in the beginning

KEY FACT

1 When the Earth's surface solidified, there were many volcanoes erupting, which gave off carbon dioxide, ammonia and methane.

Q What gas causes the greenhouse effect?

CO_2
carbon dioxide

NH_3
ammonia

CH_4
methane

KEY FACT

2 Water vapour condensed to form the oceans.

Oceans were formed on the Earth's surface from erupting volcanoes. Water vapour came out of the volcanoes. The Earth cooled and the water condensed to form liquid.

KEY FACT

3 There was little oxygen in the atmosphere.

B The effect of plants on the atmosphere

KEY FACTS

1 500 million years ago plants started to give out oxygen.

2 Methane and ammonia reacted with the oxygen.

3 Nitrifying microbes produced nitrogen from ammonia.

Q What process changes water vapour into the oceans?

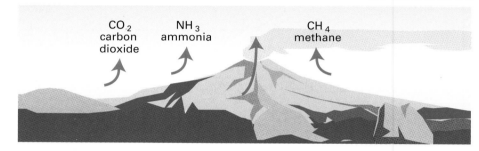

oxygen O_2
nitrogen N_2
carbon dioxide CO_2
plants
water H_2O

c   The atmosphere today

1   The composition of the air is <u>78% nitrogen</u>, <u>21% oxygen</u> and <u>1% carbon dioxide</u> and <u>noble gases</u>.

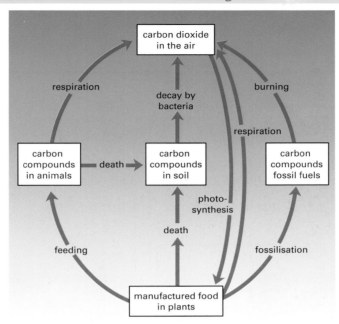

How is
rbon returned
the
tmosphere?

2   The carbon cycle helps maintain the balance of the atmosphere.

- If there is too much carbon dioxide in the atmosphere, this can lead to global warming.

3   The <u>greenhouse effect</u>: the carbon dioxide layer traps heat in the Earth's atmosphere, increasing the temperature.

The ozone layer doesn't have an effect on global warming. This is a common mistake.

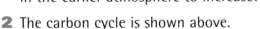

1 Explain what caused the amount of oxygen in the earlier atmosphere to increase?

2 The carbon cycle is shown above.

a) Use the diagram to find three ways in which carbon dioxide is put back into the atmosphere.

b) (i) What is photosynthesis?
   (ii) What does it do to the amount of carbon dioxide in the atmosphere?

c) Over the past 100 years, more fossil fuels have been burned.
   (i) What change might this make to the amount of carbon dioxide in the air?
   (ii) What is the change in climate that most scientists are worried this might cause?

Electric current and charge

THE BARE BONES

➤ Electric current transfers energy round a circuit.
➤ An electric current is a flow of charge, measured in amperes (A).

A **Electric current**

KEY FACT

1 Electric current is a flow of charge.

- Electric current is a **flow of charged particles** around an electric circuit. The charges **transfer energy** from a power supply to the components in the circuit.

2 Electric current is measured in **amperes (A)**.

Remember
Current is the same all the way round the circuit.

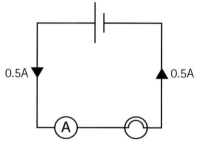

Electric current is measured with an ammeter, which is placed in the circuit **in series**.

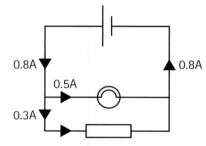

In a **parallel circuit** the currents flowing **into a junction** add up to the current flowing **out**.

Remember
You need to know where to place an ammeter and a voltmeter in the circuit.

3 Example What will the current reading be on ammeter 1 and ammeter 2?

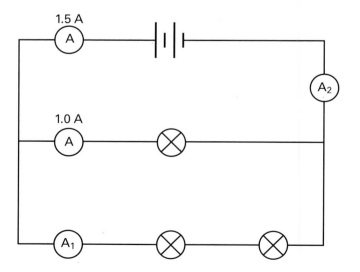

Q Where does the ammeter go in the circuit?

Current into the junction = current out of the junction, so ammeter 1 reads 0.5 A. The current returning to the cells must equal the current leaving the cells, so ammeter 2 reads 1.5 A.

B   Electric charge

1   Electrical charge (Q) is measured in coulombs (C)

- When the current is 1 ampere, a charge of 1 coulomb passes in 1 second:

  charge (coulombs) =   current (amperes)   × time (seconds)

  $$Q = It$$

2 Example The current through a lamp is 2 A. Calculate the charge that flows through the lamp in 3 minutes.

- 3 minutes = 3 × 60 seconds

  = **180 seconds**

- $Q = It$

  = 2 A × 180 s

  = 360 C

Calculations about electric charge and current are only on the Foundation tier for some specifications. Check with your teacher whether you need to know how to do them.

Can you identify all the components in the circuits on this page?

1 Complete the following sentences about electric current, voltage and resistance.

Electric current is a flow of _____. The current leaving a

power supply is _____ the current returning to the power supply.

2 The current through an electric iron is 5 A. Calculate the charge that passes through the iron when it is switched on for 10 minutes.

Voltage and resistance

THE BARE BONES
➤ Voltage tells us the difference in the energy carried by the charge between two points.

➤ Resistance tells us how difficult it is for a current to pass round the circuit.

A Voltage

1 A current only passes through a component or wire if there is a voltage across it. The bigger the voltage across a component, the bigger the current through the component.

> The higher the voltage of the supply the more energy given to the charge passing through it.

2 Voltage *V* is measured in **volts** (V).

Voltage is measured with a **voltmeter**, which is placed **in parallel** with the component, as shown below.

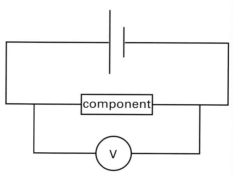

3 Example Three components are connected in this circuit. What will be the readings on voltmeter 1 and voltmeter 2?

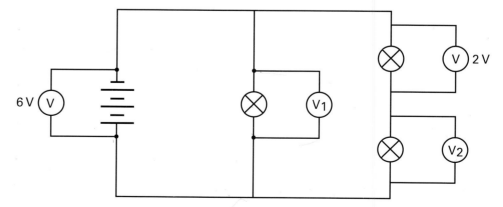

The voltage across each parallel branch is the same. So the voltage across lamp 1 must be 6 V.

The voltage across lamps 2 and 3 is a total of 6 V so there must be 4 V across lamp 2.

Q Calculate the current that passes through a cell when 180 C flows in 6 minutes.

B Resistance

1 Resistance of a wire or component tells us how difficult it is for a current to pass.

> Changing resistance in a circuit alters the current: the bigger the resistance, the smaller the current (if the voltage is the same.)

2 Most components have some resistance, but it is low in connecting wires.

3 Some wires are designed to have resistance. Resistance wires are used to make variable resistors.

* The longer the resistance wire, the bigger the resistance.

* Variable resistors can be used to control the current in a circuit. The longer the wire, the greater the resistance and the smaller the current. A lamp in the circuit will glow less brightly.

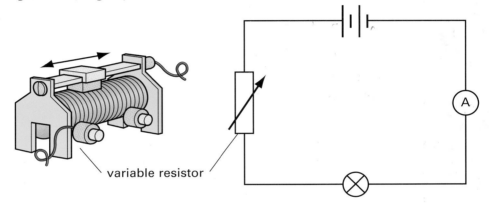

variable resistor

4 Some materials get hot when a current passes through them. The filament in a lamp is a very thin wire, which gets hot when a current passes. The same effect heats the element in an electric fire or a hairdryer.

5 Resistance *R* is measured in ohms (Ω).

6 The following equations relate voltage, current and resistance:

voltage (volts) = current (amperes) × resistance (ohms)

$$V = I R$$

7 Example The voltage across a lamp filament is 12 V when the current is 3 A. Calculate the resistance of the filament:

$$V = I R$$
$$12\,V = 3\,A \times R$$
$$R = 12\,V \div 3\,A$$
$$R = 4\,\Omega$$

emember
ou need to be
ble to recall and
se the equation
= IR.

What is the
oltage across a
0 Ω resistor
hen the
urrent is 2 A?

1 Complete the following sentences:

The voltage across a component tells us how much _____ is transferred to the component. The shorter the wire the _____ the resistance and the _____ the current. A lamp in the circuit will glow _____.

THE BARE BONES

➤ The electric current in a circuit is controlled by the voltage of the supply and the resistance of the components in the circuit.

➤ Current-voltage graphs are used to show how the current through a component depends on the voltage across it.

A Resistors

1 A resistor is a component that is designed to have a constant resistance.

The electric current through a component depends on the voltage across it.

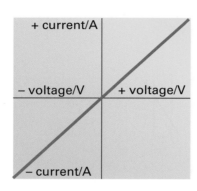

A resistor at a constant temperature has constant resistance.

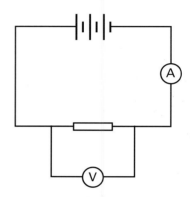

Resistance = voltage / current. The graph is a straight line so V / I is constant.

Remember
You need to know the symbols for a resistor, lamp, cell, voltmeter and ammeter.

Example: Copy the current-voltage graph for the resistor and draw a second line to show the how the current changes with voltage across a resistor with twice the resistance.

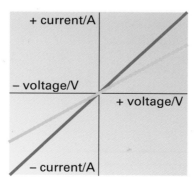

The resistance is twice as big, so at any voltage the current is half as much.

Q What is the equation linking V, I and R?

B   Lamps

1 As the voltage increases across a lamp's filament it gets hotter, increasing its resistance.

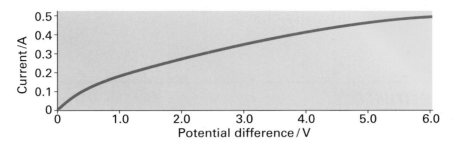

As the voltage increases, the current does not increase as much because the resistance has increased.

> The resistance of metals increases as the temperature increases.

2 The electric current that makes a wire hot heats many everyday electrical appliances, e.g. iron, hairdryer, tumble dryer and electric kettle.

How does the resistance of wire change as it gets hot?

EY FACT

C   Diodes

1 Diodes let current pass one way only. If the voltage is reversed no current can pass.

2 The diode in the circuit below allows the current to pass through it. If the diode is reversed lamp A will light normally, but lamp B will not light because the diode does not allow current to pass through that branch of the circuit.

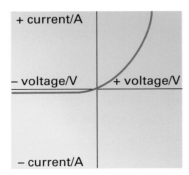

No current flows when the voltage is reversed.

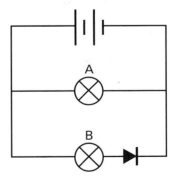

Why is a diode sometimes called a valve?

PRACTICE

1 Complete this sentence:
The resistance of a lamp filament increases as the temperature _____ .

THE BARE BONES

➤ Thermistors are used to monitor changes in temperature.

➤ The resistance of an LDR (light-dependent resistor) changes as the light intensity changes.

A   Thermistors

1 A **thermistor** is a semiconductor component that can be used to detect changes in temperature.

KEY FACT

> The resistance of a thermistor gets less as its temperature increases.

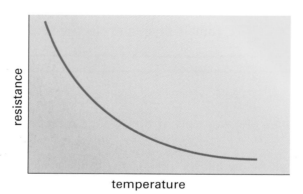

The resistance-temperature graph shows that the resistance of the thermistor decreases as the temperature rises. If the resistance in the circuit decreases the current will increase.

2 Example A thermistor is connected up in the circuit shown below. The thermistor is immersed in a beaker of water. Describe how the current in the circuit will change as the water is warmed up.

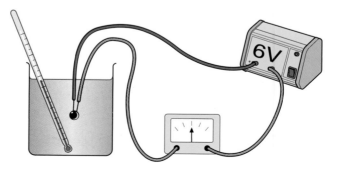

Q. Can you draw the circuit diagram for the thermistor circuit shown here?

B Light-dependent resistors (LDRs)

1 An LDR (light-dependent resistor) can be used to detect changes in light intensity.

> The resistance of an LDR gets less as the light level increases.

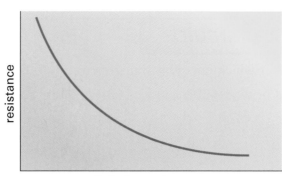

light intensity

The resistance–light intensity graph shows that the resistance of the LDR decreases as the light intensity rises. If the resistance in the circuit decreases the current will increase.

2 Example An LDR is connected up in the circuit shown below. The LDR is moved towards the lamp. Describe how the current in the circuit will change as the LDR gets closer to the lamp.

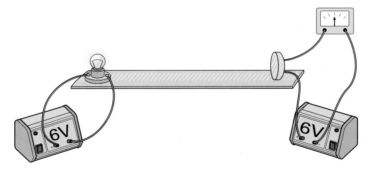

Can you
aw and label
e symbols for
resistor, a
ode, an LDR
nd a
ermistor?

1 Draw the circuit diagram for the LDR circuit shown in section B above.

2 Complete these sentences:

The resistance of an LDR increases as the light level _____ .
The resistance of a thermistor increases as the temperature

_____ .

Electricity, energy and power

THE BARE BONES

➤ Electric current transfers energy from a source to the circuit.
➤ The rate at which energy is transferred to the circuit is the power.
➤ Efficiency is the proportion of useful power transferred in a process.

A Power in electric circuits

1 Power is the rate at which energy is transferred.

• Power is measured in watts (W) and kilowatts (kW): 1 kW = 1000 W.

KEY FACT

> electrical power (watts) = voltage (volts) × current (amps)
> $P = V I$

2 Example Philip uses an immersion heater to heat water. The voltmeter reads 12 V and the ammeter reads 3 A. What is the power transferred to the heater?
power (W) = voltage (V) × current (A)
power = 12 V × 3 A = 36 W

> Write down the equation you'll use before you begin a calculation.

B Energy in electric circuits

1 Batteries, solar cells and generators are all sources of energy. Electric current transfers the energy from the source to the circuit.

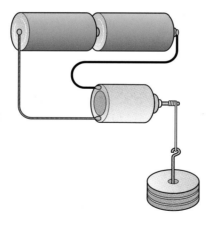

Energy from the chemical reactions in the battery makes the motor turn and lift the load.

The amount of energy transferred to an electrical appliance depends on the time it is switched on and the power-rating of the appliance.

Remember
You need to be able to recall and use $P = V I$.

KEY FACT

> energy transferred (joules) = power (watts) × time (seconds)

2 Example How much energy does a 100 W immersion heater transfer in 5 minutes?

60 seconds in 1 minute so 5 minutes = 300 seconds

energy transferred (joules) = power (watts) × time (seconds)

energy transferred = 100 W × 300 s = 30000 J

Q What device uses a chemical reaction to generate electricity?

C Energy in the home

1 The electricity meter measures the energy supplied by the electricity company.

- It measures energy in units called kilowatt-hours (kWh).
- One kWh is the energy transferred when 1000 watts is transferred for one hour.

We can calculate the energy transferred by an appliance using:
energy (kWh) = power (kW) × time (hours)

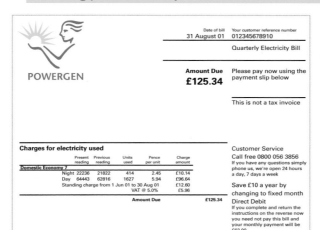

2 Example Electricity costs 6 pence per unit (1 kWh). What is the cost of running a 3 kW electric heater that is switched on for 4 hours?

energy transferred in 4 hours = 3 kW × 4 hours = 12 kWh
cost = 12 kWh × 6 pence per kWh = 72 pence

How many units does a kW motor use 3 hours?

D Efficiency

1 Efficiency tells us how much of the power input becomes useful output.

$$efficiency = \frac{useful\ power\ output}{power\ input} \times 100\%$$

2 Example A 40 W filament lamp gets very hot. In fact 90% of the power transferred by the lamp warms up the room rather than lighting it. How much power is transferred to lighting the room?

10% of the power from the lamp is useful to light the room. 10% of 40W is 4 W.

You need to be able to use the equation for efficiency.

What is the efficiency of a kW motor that produces .5 kW of useful power?

PRACTICE

1 A current of 3 A passes through a 12 V lamp. Calculate the rate at which the lamp transfers energy from the supply.

2 A 12 W lamp is switched on for 10 minutes. Calculate how much energy is transferred from the power supply.

3 A 15 W lamp transfers 6 W of its energy to light. Calculate the lamp's efficiency.

Electricity at home

> The current from batteries is a direct current (dc).
> Mains electricity supplies an alternating current (ac).
> The electric current supplies energy to the house through the live wire and returns through the neutral wire.

A Direct current and alternating current

1 **Batteries** produce **direct current (dc)**.

- Direct current flows in the same direction
- Direct current does not change in size.

2 **Mains electricity** produces **alternating current (ac)**.

- Alternating current is constantly changing size and direction.

Example What is the frequency of the alternating current in the graph below?

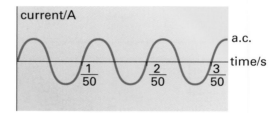

The direction changes 50 times per second – the frequency is 50 Hz.

Q What sort of current is supplied by a battery?

B Keeping safe

1 Power cables have three strands – live, neutral and earth:

- live – supplies energy to the house
- neutral – energy supply leaves the house
- earth – this wire does not normally carry a current.

2 The strands are insulated so no current passes between the wires.

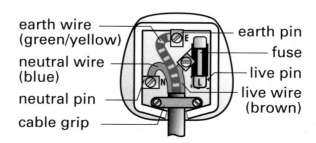

earth wire (green/yellow) — earth pin
neutral wire (blue) — fuse
neutral pin — live pin
cable grip — live wire (brown)

B

3 Fuses and circuit breakers are designed to break the circuit if the current is too high.

You should be able to explain how a fuse and circuit breaker protect the circuit.

4 The fuse in a plug is a thin resistance wire.

- the fuse melts if the current gets too large.
- the circuit breaks and prevents damage to the appliance.
- the fuse must match the current rating of the appliance.

5 A fuse or circuit breaker near the electricity meter will break the circuit if the current is too large. This stops the house wiring overheating and prevents a fire.

Why should the fuse rating be as close to the appliance current rating as possible?

- The metal strings of the guitar are connected to the earth pin on the plug.
- If there is a fault and the strings become live a large current might pass through the guitarist to reach earth.
- The large current will trip the circuit breaker.
- A circuit breaker will cut the circuit more quickly than a fuse.

6 Circuit breakers give better protection than fuses.

7 Some appliances, e.g. hairdryers, are marked as having double insulation (symbol on the right). There are no electrical connections to the casing of the appliance and no earth connection.

PRACTICE

1 Fill in the gaps in the table using words from this list:
metal, plastic, insulator, conductor, low melting point, high melting point

component	property	material
inner core of cable		copper
outer covering of cable		
fuse		thin resistance wire

2 Calculate the current that normally passes through a kettle rated 2 kW and connected to a 230 V supply. Should the plug have a 3 A fuse or a 13 A fuse? Explain your answer.

3 Describe the difference between direct current and alternating current.

Electric charge

➤ Electrons have a negative charge.

➤ When some materials are rubbed together they become charged.

➤ Like charges repel each other. Opposite charges attract each other.

➤ A charged object often loses its charge with a spark.

A Electrostatic charge

1 When some materials are rubbed together they become charged. Electrons have been transferred from one material to the other.

2 If the materials are insulators the charge does not leak away. This is sometimes called static charge.

3 The object that loses electrons will be positively charged. The object that gains electrons will be negatively charged.

KEY FACT

4 Like charges repel each other. Opposite charges attract each other.

Charging your hair with a hairbrush makes it stand on end.

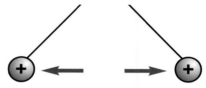

Two objects with the SAME charge REPEL each other.

Two objects with DIFFERENT charge ATTRACT each other.

KEY FACT

5 When a charged object discharges it may do so with a spark – which can be dangerous.

Petrol passing along the refuelling pipe causes enough friction to give the pipe an electrostatic charge. If the charge builds up there is the danger of sparking and risk of an explosion. The pipe is connected to earth, so that the charge can leak away safely.

6 Example When you take off a jumper in the dark you might see and hear sparks. Explain why.

Pulling your jumper over your head charges up the jumper and your head. As the opposite charges are attracted to each other, they jump across the air gaps, creating a small electric current – a spark.

Q Can you explain why your hair stands on end and 'crackles'?

B Using electric charge

A photocopier and a laser printer both use electrostatic charge to print on paper.

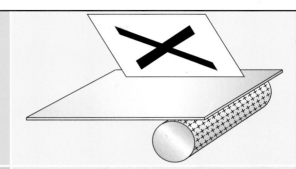

- The page to be copied is placed face down on a sheet of glass.
- The surface of the drum is coated in a material that emits electrons when light falls on it.
- When the copier is switched on the surface of the drum becomes positively charged.

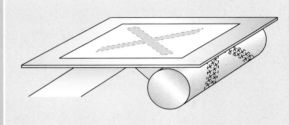

- A bright beam of light moves across the page. Light is reflected from white areas of the paper and reflected onto the drum below.
- Wherever light hits, electrons are emitted from the material of the drum and neutralise the positive charges. Dark areas on the page do not reflect light onto the drum, leaving a pattern of positive charges on the drum's surface.

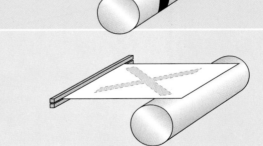

- Negatively-charged, dry, black powder called toner is dusted over the surface of the drum, and the pigment particles are attracted to the positive charges.

- A positively-charged sheet of paper then passes over the surface of the drum, attracting the beads of toner away from it.
- The paper is then heated and pressed to bond the image formed by the toner to the paper's surface.

Remember
Like charges repel. Opposite charges attract.

Q Can you explain why the TV screen gets very dusty quickly?

Example: Electric charge is used to remove ash from the smoke in a coal-fired power station. Long metal electrodes are fitted inside the chimney. The electrodes are connected to a high voltage supply. Explain how this will clean up the smoke.

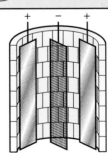

The ash particles in the smoke have positive and negative ions attached. The particles are attracted to electrodes in the chimney.

PRACTICE

1 A wire links a petrol tanker to the earth when the tanker is delivering fuel. Explain how this helps prevents a potential explosion.

2 Julie walks across a wool carpet and then touches a metal rail. She feels a small electric shock. Explain why.

Electromagnetism

➤ There is a magnetic field around a magnet.

➤ A magnetic field is created around a wire when a current flows in it.

➤ A force acts on a wire carrying an electric current through a magnetic field.

A Magnetic fields

1 The magnetic field around a magnet affects other magnets and also magnetic materials, e.g. iron and nickel.

• Field lines show the direction of forces around the magnet.

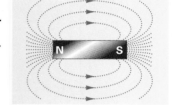

KEY FACT

2 The field near a single wire carrying an electric current is circular.

3 The field near a coil carrying an electric current looks very similar to the field near a bar magnet.

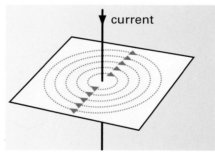

current

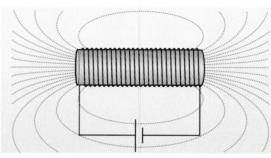

We can make the field near a coil stronger by:

• putting an iron bar through the middle of the coil
• increasing the current in the coil
• increasing the number of turns in the coil.

> Each separate change in direction – of current or field, will change the direction of the force.

Q How can you make the magnetic field stronger?

B Magnetic forces

KEY FACT

When a current-carrying wire is at right angles to a magnetic field there is a force on the wire.

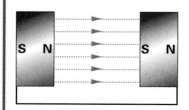

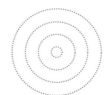

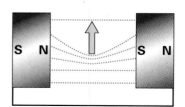

When the wire is in a magnetic field, the two fields combine to produce a 'catapult' field which pushes the wire upwards.

Q How can you make the catapult force bigger?

c Electric motor

A simple electric motor has a coil of wire in a magnetic field. When a current flows the motor spins.

emember

the magnetic
ld and the
rrent are at
ht angles,
e wire jumps
the third
ection.

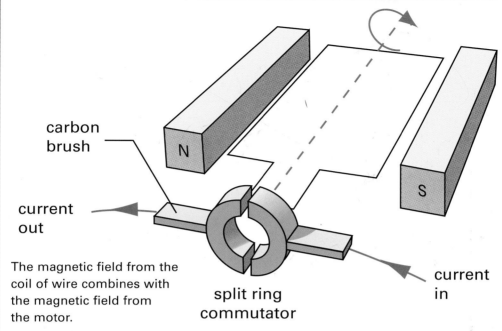

carbon brush

current out

split ring commutator

current in

The magnetic field from the coil of wire combines with the magnetic field from the motor.

The catapult fields push the right side of the motor up and the left side down.

The coil begins to turn.

The coil turns over so that the wires are in contact with the brushes again and the right side is again pushed up. The split ring commutator makes sure the current always passes into the side of the coil on the right – making the current reverse its direction inside the coil.

How can you ake the motor in faster?

Example: Suggest two ways to change the direction in which the coil spins.

Answer: Reversing the current or magnetic field would reverse the direction of spin.

RACTICE

1 When a current passes through a wire there is a _____ magnetic _____ in the space around the wire. If the current-carrying wire is at right angles to a magnetic field there is a _____ on the wire at right angles to both the magnetic field and the wire. The size of the force depends on the size of the _____ and the strength of the _____ .

2 Complete the diagram with labels to show the direction of:
   a) the current through the coils
   b) the magnetic field due to the permanent magnets
   c) the force on the coil.

Electromagnetic induction

THE BARE BONES

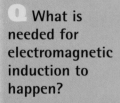

➤ Spinning electromagnets near coils of wire generates electricity.
➤ Transformers need alternating current (ac) to use electromagnetic induction.

A Electromagnetic induction

KEY FACT

1 Moving a magnet into a coil produces a voltage across the ends of the coil.

When the magnet is pulled out the voltage is reversed. If the magnet is stationary there is no voltage.

2 This way of generating a voltage is called **electromagnetic induction.**

3 Electricity can be generated by rotating a magnet inside a coil or by rotating a coil in a magnetic field.

4 **Example** Suggest two ways in which the voltage induced across the coil could be increased.

Answer The voltage increases as the speed at which the coil is moved increases. The voltage is also greater if a stronger magnet is used.

Q What is needed for electromagnetic induction to happen?

B Generators

KEY FACT

1 A generator consists of a coil spinning inside a magnetic field or a magnet spinning between coils.

2 A cycle dynamo is driven by the wheel of the bike.

The dynamo spins a magnet inside an iron core. A changing magnetic field in the iron core induces a voltage across the coil.

magnet rotates

coil

iron core

3 **Example** Suggest three ways in which the voltage output from the generator could be increased.

Answer The voltage can be increased by:
- spinning the coil faster
- using a stronger magnet
- having more turns of wire on the coil.

Q What would be the effect of spinning the dynamo the opposite way?

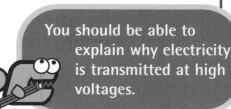

You should be able to explain why electricity is transmitted at high voltages.

C Transformers

1 A transformer has an iron core and two coils of wire.

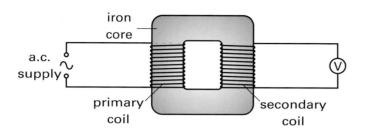

iron core

a.c. supply

V

primary coil

secondary coil

2 The primary coil is connected to an alternating current supply.

3 As the current in the primary coil varies it sets up a changing magnetic field in the iron core.

4 The changing field in the iron core induces a changing voltage in the secondary coil.

5 The voltage in the secondary coil depends on the number of turns on the coil.

6 The bigger the number of turns on the secondary coil, the bigger the voltage across the coil.

7 A transformer needs alternating current to create a changing magnetic field to induce a voltage in the secondary coil.

Example A transformer has 200 turns on the primary coil and 400 turns on the secondary coil. Explain why this transformer is called a step-up transformer.

Answer There are more turns on the secondary coil so the output voltage will be greater than the input voltage. The transformer has increased the voltage. This is a step-up transformer.

Remember
There has to be a changing magnetic field near a wire or a wire moving in a magnetic field for electromagnetic induction to happen.

KEY FACT

Why does a transformer need an alternating current?

PRACTICE

1 When the north end of a magnet is pushed into the right hand side of a coil, the meter needle flicks to the right.

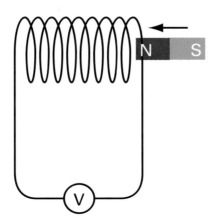

N   S

V

Explain what happens when each of the following occurs:
(a) the north end of the magnet is pulled out of the coil
(b) the north end of the magnet is pushed in the left-hand end of the coil
(c) the south end of the magnet is pushed into the right-hand side of a coil
(d) the south end of the magnet remains stationary in the coil.

Distance, speed and acceleration

THE BARE BONES
- ➤ The speed of a moving object tells us the rate at which it moves.
- ➤ The velocity of a moving object tells us the speed and direction.
- ➤ Acceleration tells us how much the velocity changes each second.
- ➤ Distance-time and speed-time graphs can describe a journey.

A Distance and speed

KEY FACT

1 If an object moves in a straight line, the average <u>speed</u> of the object can be worked out.

$$\text{speed (m/s)} = \frac{\text{distance travelled (m)}}{\text{time taken (s)}}$$

Remember
When calculating speeds use time in seconds.

2 A distance-time graph (see below) shows how far a car goes in a certain time. The gradient of the slope tells us whether the car is moving quickly or slowly.

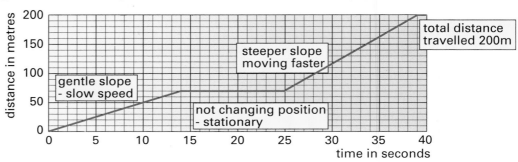

total distance travelled 200m

steeper slope moving faster

gentle slope - slow speed

not changing position - stationary

distance in metres (vertical axis: 0, 50, 100, 150, 200)
time in seconds (horizontal axis: 0, 5, 10, 15, 20, 25, 30, 35, 40)

Q How far would a car moving at 30 m/s travel in 20 seconds?

3 Example The average speed of a boy who travels 300 m on a bike in one minute:

Answer
$$\text{speed (m/s)} = \frac{\text{distance travelled (m)}}{\text{time taken (s)}} = \frac{300m}{60s} = 5m/s$$

B Velocity

KEY FACT

The <u>velocity</u> of a moving object describes the speed and direction.

Example A girl swings a ball on a string around her head at a constant speed of 3 m/s. Use a diagram to explain how the velocity changes.

(a) When the space shuttle is being put into orbit, it is crucial that the scientists at NASA know how fast it is going and in which direction.

(b) When the space shuttle is in orbit around the Earth it is travelling at a constant speed, but it is changing direction all the time because it is pulled by the Earth's gravity. Its velocity is constantly changing.

Q What is the difference between speed and velocity?

C Acceleration

1 Acceleration tells us how much the velocity changes each second.

$$\text{acceleration (m/s}^2) = \frac{\text{change in velocity (m/s)}}{\text{time taken (s)}}$$

2 A velocity-time graph tells a story of how the velocity changes during a journey.

The gradient of this graph tells us about the acceleration of the bus.

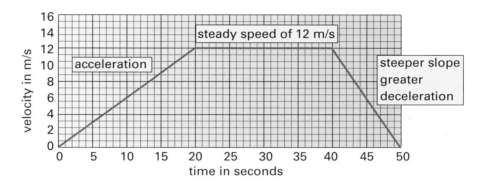

steady speed of 12 m/s

acceleration

steeper slope greater deceleration

3 Example During lift-off the shuttle accelerates from rest to 8400 m/s in 8 minutes. Calculate the acceleration.

$$\text{acceleration (m/s}^2) = \frac{\text{change in velocity (m/s)}}{\text{time taken (s)}}$$

$$\text{acceleration (m/s}^2) = \frac{8400 \text{ m/s}}{480 \text{ s}} = 17.5 \text{m/s}^2$$

Look carefully at the labels on the axes when interpreting graphs.

What would feel like to be ccelerating at .5 m/s² for 8 inutes?

---

1 A boy runs 400 m in 80 s. What is his average speed?

2 Look at the velocity-time graph for a bus, below. Describe what is happening in each part of the journey, O–A, A–B, B–C, C–D, D–E, E–F and F–G.

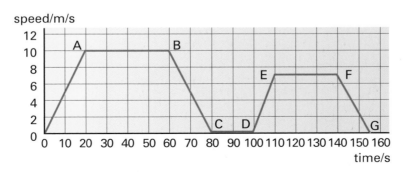

3 A mini car starts a race by accelerating from 0 to 30 m/s in 10 s. What is the acceleration of the car?

Forces - in pairs and balanced

➤ Force is measured in newtons (N).

➤ When two bodies interact the forces exerted on each other are equal and opposite.

➤ Equal, balanced forces acting on an object will not enable movement.

A   Forces come in pairs

KEY FACT

1   Whenever two objects interact they exert equal and opposite forces on each other.

- When you sit on a chair you compress the springs in the cushion. The force from the springs pushes up on you as your weight pushes down to compress the springs.

- If the chair did not push back you would fall through the chair on to the floor!

2 Example What is the force that makes a sprinter accelerate away from the starting blocks?

- The sprinter pushes back on the block.

- The bonds between the atoms in the block are compressed – the atoms get closer together, just like the springs in the cushion.

- The particles in the block push forward on the sprinter – the harder he pushes back – the harder the block pushes forwards.

Q What can you say about the forces acting on an object moving at a constant speed?

B Balanced forces

emember

forces
agrams the
ze of the force
row is related
the size of
e force.

If the forces on an object are balanced there will be no effect on the movement of an object.

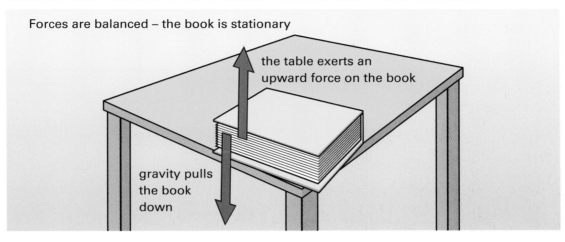

Forces are balanced – the book is stationary

the table exerts an upward force on the book

gravity pulls the book down

Example What are the balanced forces acting on a bicycle moving at a steady speed along the road?

The driving force of the cyclist's legs are balanced by the drag forces against the motion. (Read more about drag forces on page 134)

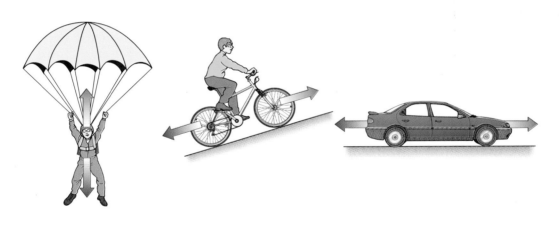

What are
e forces that
t on a book
sting on a
ble?

1 Each of the items above is moving at a constant speed. What are the forces acting on them?

Forces - mass and acceleration

THE BARE BONES
> An unbalanced force on an object causes it to accelerate.
> The bigger the unbalanced force on an object, the bigger the acceleration.

A Unbalanced forces

KEY FACT

An unbalanced force changes the velocity of an object.

1 If the forces on an object are unbalanced, it will change the velocity of the object:

Remember
No force – no acceleration.

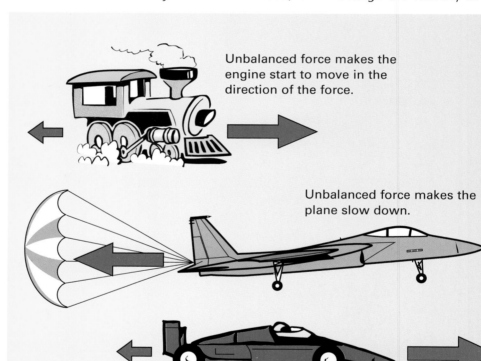

Unbalanced force makes the engine start to move in the direction of the force.

Unbalanced force makes the plane slow down.

Unbalanced force makes the car increase speed.

2 Example Describe the direction and size of the horizontal forces that act on a full shopping trolley rolling to a stop, without being pushed.

Answer The only forces acting on the trolley are drag forces slowing it down.

Q What can you say about the forces acting on an object that is accelerating?

B   Force, mass and acceleration

1   When an unbalanced force makes an object accelerate:

- the bigger the force, the bigger the acceleration
- the bigger the mass, the smaller the acceleration.

- A car with a bigger engine will accelerate faster than a car with a smaller engine.

- Someone pushing a wheelbarrow full of leaves will accelerate faster than someone pushing a wheelbarrow full of stones.

2 Example A student is cycling home from school with a bag full of books on the carrier. Explain why it is harder to get started from the traffic lights when he has a bag of books, than when his carrier is empty.

Answer The bag of books increases the mass of the bike. A larger mass makes it more difficult to accelerate.

When answering questions about forces, draw a diagram showing force arrows to help you to understand the situation.

1 Each of the pictures below has two arrows to show the forces on the object. Write down whether you think the object is speeding up, slowing down or stationary, and why.

Stopping forces

THE BARE BONES

➤ Resistive forces act in the opposite direction to the movement of an object.

➤ Thinking distance depends on the speed of a car and the reactions of the driver.

A Resistive forces

KEY FACT

1 Resistive forces are the forces that resist the movement of an object over a surface or through a liquid or a gas. They are also called frictional (or drag) forces.

Resistive forces often result in objects warming up as they are slowed down. The space shuttle has insulation to prevent the inside of the orbiter becoming overheated as the spacecraft returns through the atmosphere to Earth. See page 138 for more about air resistance.

2 There are many resistive forces in a car.

• Friction between the tyres and the road is needed to drive the car along.

• Friction between the brake shoes and the brake pads slows down the car.

• Friction between moving parts in the engine.

• Air resistance slows the car down. Cars are designed to be streamlined to reduce the air resistance.

Remember
The time a car takes to decelerate will depend on the forces acting and the mass of the car.

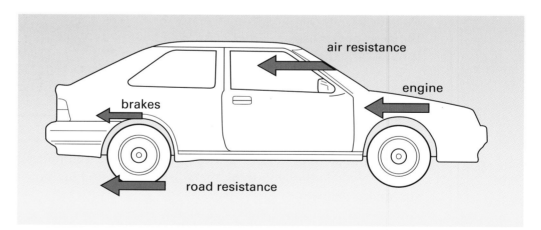

3 Example A Grand Prix car uses smooth tyres called 'slicks' on a dry track. The tyre softens as it gets warm with running. Explain why the tyre might be good on a dry track but no good on a wet track.

Answer The soft tyre sticks well to the dry track, creating good grip. On a wet track the smooth tyre would skid across the surface. A grooved tyre provides better grip – the water escapes through the groove, giving better contact with the road.

Q Where are the resistive forces that slow a car down?

B Thinking distance

Y FACT

1 Thinking distance is the distance the car travels between the driver realising he has to put the brakes on and the brakes actually going on.

2 The thinking distance depends on the reaction time of the driver. The reaction time of the driver is affected by:

- tiredness – sleepiness is a common cause of motorway accidents
- drugs or alcohol – drugs and alcohol slow down a driver's reactions
- poor visibility – a foggy day makes it harder to see what is ahead
- distractions in the car – such as talking on a mobile phone or noisy children.

Example Why is it dangerous to drive after drinking alcohol or taking drugs?

Drugs or alcohol make the driver less alert – he will take longer to notice a hazard and to apply the brakes. While he is reacting the car will still be travelling towards the hazard.

Y FACT

3 The thinking distance also depends on the speed of the car.

- A car travelling at 30 m/s travels twice as far as a car travelling at 15 m/s before the driver applies the brakes in an emergency.

What are e factors that ould slow wn a driver's actions?

RACTICE

1 Objects slowed down by resistive forces often become warmer. Where has the energy come from to make them warm?

2 What is the effect on the thinking distance if the driver is tired?

Stopping a car

THE BARE BONES

➤ The distance a car travels while the brakes are applied depends on the braking force, mass of the car and its occupants, and its speed.

➤ The stopping distance depends on the speed of the car, the driver's reactions, braking distance and condition of the car and road.

A Braking distance

KEY FACT

1 The <u>braking distance</u> of a car is the distance it travels between the brakes being applied and the car coming to a halt.

The **mass of the vehicle** is important – the more massive the car, the greater the braking distance.

The **speed of the car** is important – the faster the car is going, the greater the braking distance.

The **force from the brakes** is important – the greater the force from the brakes, the shorter the braking distance. Poorly maintained brakes make it harder to stop.

The **road conditions** are important – on a wet road there will be less friction between the tyres and the road – the braking distance is longer. If the driver brakes too hard the car may skid.

Example Explain why a lorry loaded with bricks will take longer to stop than an empty lorry.

Answer The lorry loaded with bricks has a much greater mass.

It will have a smaller deceleration for the same force – so it will travel further before stopping.

Q What are the factors that affect the braking distance of a car?

You should be able to describe the range of factors that affect car stopping distances.

B Stopping

Y FACT

1 The overall stopping distance of a car depends on the time it takes for the driver to react and the distance the car travels while braking.

overall stopping distance = thinking distance + braking distance

2 Look at the picture above. It is a foggy day. The driver of the truck is tired and cannot see well in the fog. Suddenly he applies the brakes – but, too late, he runs into the car in front.

When the truck runs in to the red car in front both will be damaged. The truck exerts a force on the red car. The red car exerts an equal and opposite force on the truck. This is another example of forces coming in pairs.

3 **Example** John drives a lot of miles on motorways in his job. When he takes the family on holiday there are three extra people and all their luggage. Why must he remember to allow more braking distance?

Answer The car is now much heavier – more mass will take longer to stop, so he must allow extra distance between his car and the one in front.

Q What are the factors that affect the overall stopping distance of a car?

PRACTICE

1 Write down what effect each of the following situations has on the stopping distance of a car and explain why.
a) The car is heavily loaded.
b) The car is moving quickly.
c) The road is wet.

2 A car is travelling at 30 m/s. The driver has a reaction time of 0.5 s.
a) Calculate the thinking distance - how far the car travels between the driver seeing he needs to brake and applying the brakes.
b) What would be the effect on this thinking distance if the driver were tired? Explain your answer.
The braking distance for this car to brake from 30 to 0 m/s is 64 m.
c) What is the total stopping distance?

THE BARE BONES
➤ Objects fall because gravity pulls them.
➤ When objects fall through air, there is also air resistance acting on them.

A   Gravity

1 A gravitational force is felt near any large, massive object, such as the Sun, Moon, Earth or other planets *(see page 152)*.

2 Gravity is the force that pulls falling objects to the ground.

3 The force of gravity on an object (its weight, measured in newtons) depends on two things:

- the mass of the object, m
- the strength of the gravitational field, g.

KEY FACT

> weight (N) = mass (kg) × gravitational field strength (N/kg)
>
> $$W = mg$$

4 It is important to appreciate the difference between the mass of an object, measured in kilograms and the weight of an object, a force measured in newtons.

5 At the surface of the Earth the gravitational field strength (g) is approximately 10 N/kg. This means that the weight of 1 kilogram of matter is 10 newtons.

6 Example Anna weighs 60 kg. What is her weight in newtons?

$$W = mg$$
$$= 60 \text{ kg} \times 10 \text{ N/kg} = 600 \text{ N}$$

Q What is your weight in newtons on the Earth?

B   Air resistance

1 When objects fall through air, air resistance acts on them.

KEY FACT

> The amount of air resistance depends on how fast the object is falling and its shape.

2 The bigger the surface area, the greater the air resistance.

3 The faster the object moves, the greater the air resistance.

Example Car designers try to design cars with low air resistance so they need a smaller driving force, using less fuel.

Q What are the factors that affect the air resistance of a moving object?

C Terminal velocity

1 Reaching terminal velocity

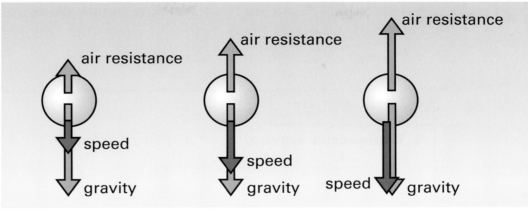

- As a falling object speeds up, its air resistance increases.

- Eventually, if it falls fast enough, the pull of gravity (its weight) is balanced by the air resistance.

- It stops accelerating and falls at a steady speed – called its terminal velocity.

EY FACT

2 The bigger the weight of the object, the faster it has to be falling to reach terminal velocity.

3 Example The space shuttle releases its booster rockets over the sea to be retrieved. How does the speed of the booster rocket change after its release?

Answer At first the rocket accelerates. When the parachute opens, the air resistance increases and it reaches a steady speed.

What are the factors that affect the terminal velocity of a falling object?

Use your ideas about balanced forces, weight and mass to help your explanations about terminal velocity.

PRACTICE

1 A parachute acts to increase the surface area of a falling object.

a) How do the forces on the parachutist change as she falls?
b) What happens to her speed?

2 Explain why aircraft landing on an aircraft carrier use a 'drag chute', but they do not use one when landing on an airfield.

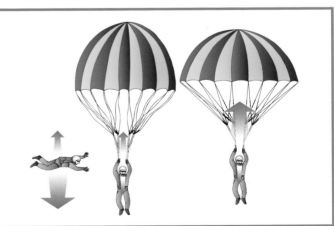

Properties of waves

THE BARE BONES
- ➤ Waves transfer energy without transferring matter.
- ➤ Wave oscillations are in different orientations, depending on the type of wave and the direction in which the energy travels.
- ➤ Waves can be reflected and refracted.

A   Making waves

1 Transverse waves

> **KEY FACT** ▶ In transverse waves, the <u>oscillation is at right angles</u> to the direction in which the energy travels.

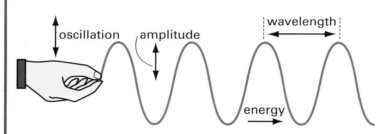

When one end of a rope is moved up and down the rest of the rope begins to move up and down and energy is carried along the rope by a wave.

Waves along a rope, light waves and waves on water are examples of transverse waves.

2 Longitudinal waves

> **KEY FACT** ▶ In longitudinal waves, <u>the oscillation is in the same direction</u> as the direction the energy is carried.

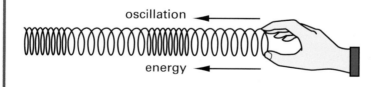

When a slinky spring is pushed backwards and forwards the rest of the spring moves in the same way. Sound is also a longitudinal wave.

The frequency of a wave is the number of waves produced each second. The unit of frequency is hertz (Hz).

> **KEY FACT** ▶ **3**   the equation relating speed, frequency and wavelength:
> wave speed (m/s) = frequency (Hz) × wavelength (m)
> $v = f \lambda$

Example A musical note has a frequency of 440 Hz. The wavelength of the sound is 0.75 m. Calculate the speed of sound.

$v = f \lambda = 440 \text{ Hz} \times 0.75 \text{ m} = 330 \text{ m/s}$

Q What happens to a cork floating on water when a wave passes?

B Reflection

1 When a wave hits a barrier it **reflects** (bounces back).

- The waves reflected from the barrier make the same angle with the barrier as the incoming waves.

- The wavelength and speed of the wave remain the same.

2 The same effect occurs when light is reflected from a mirror.

angle of reflection = angle of incidence.

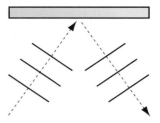

What is the rule linking the angle of incidence and the angle of reflection?

C Refraction

1 When water waves pass into shallower water they are slowed down. This change in speed makes the wave change direction – they are **refracted**.

2 We see the same effect when light passes from air into glass or water. The speed of light in glass is less than in air.

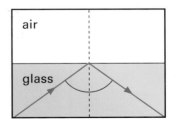

shallow water

deeper water

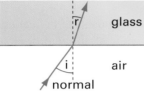

- Light is refracted towards the normal as it is slowed down.

- This time light is refracted away from the normal as it speeds up.

Use small arrows to show the direction of light rays.

Why does refraction occur?

D Total internal reflection

1 When light reaches the boundary between glass and air, the light emerging into the air is refracted away from the normal.

2 If the angle of incidence is large enough, the light is reflected back inside the block.
This is total internal reflection.

air

glass

In which direction is light refracted when it goes from glass to air?

PRACTICE

1 Complete the ray diagram to show how total internal reflection is used to make these two prisms into a periscope.

Electromagnetic spectrum

THE BARE BONES

➤ The electromagnetic spectrum is a family of waves that all travel at the same speed in a vacuum.

➤ Electromagnetic waves are transverse waves.

➤ Electromagnetic waves can travel through a vacuum.

A The electromagnetic spectrum

short wavelength 0.000 000 000 001 m high frequency Hz highest energy	gamma rays	• emitted by some radioactive materials (*see page 162*) • used in medicine to kill cancer cells 10^{20} and to trace blood flow • used to kill harmful bacteria on food and surgical instruments • large doses may damage human cells
	x-rays	• pass through soft tissue but not bones or metals • used to kill cancer cells and to produce shadow images of bones • large doses may damage human cells
	ultraviolet	• causes tanning of the skin • large doses may cause skin cancer
wavelength 0.000 000 5 m	violet	• detected by the eye
	visible light	• used for vision and photography • used through optical fibres for viewing inside the human body and other inaccessible places
	red	
	infrared	• is radiated from warm and hot objects and causes heating • used in grills, toasters, radiant heaters • used in remote control devices, security cameras and communication
	microwaves	• some wavelengths absorbed by water; used in cooking; can damage living cells • longer wavelengths used in radar, mobile phones and satellite communication
long wavelengths 1 m –100 000 m low frequency (104 Hz) lowest energy	radio waves	• broad range of wavelengths used in communication • radar used to track aircraft and shipping • radar guns measure the speed of cricket balls and cars

Q Suggest why gamma rays are more dangerous than visible light.

B Uses and hazards of the electromagnetic spectrum

1 Short wavelength electromagnetic waves, such as x-rays, gamma rays and ultraviolet have very high energies.

These waves can damage living cells. X-rays and gamma rays are used to kill cancer cells. The radiation must be applied carefully so healthy cells are not damaged.

2 Total internal reflection is used to send light rays down optical fibres.

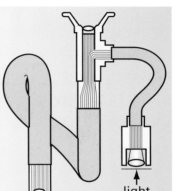

total internal reflection

- An optical fibre is a thin, glass strand with a protective coating. Light reflects off the strand's inner surfaces, so when the fibre bends, the light stays in the fibre.

- An endoscope is a device used by doctors to see inside patients without the need for operating.

3 All warm bodies emit infrared radiation.

- Infrared security cameras are sensitive to infrared radiation. They can be used to 'see in the dark'.

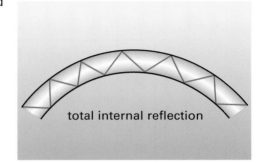

light

4 Microwave ovens produce electromagnetic waves with a wavelength around 10 cm.

- Water, fats and sugars easily absorb these waves.

- The molecules vibrate more vigorously, making the food in the oven hot.

- Microwaves could also cook you! The door of the oven reflects the microwaves back inside and safety locks prevent the oven operating whilst the door is open.

5 **Example** Light from the Sun includes ultraviolet waves. Explain why it is essential to limit the time spent exposed to ultraviolet light.

Answer Ultraviolet waves have high energy that can cause damage to skin cells, leading to skin cancer.

emember
ong wavelength
low frequency
low energy;
hort
avelength –
gh frequency –
gh energy.

Why can
xposure to too
any x-rays be
angerous?

PRACTICE

1 Fill in the missing waves in the gaps in this electromagnetic spectrum:
gamma rays → _____ → ultraviolet → visible → _____
→ microwaves → _____

2 Suggest one use and one hazard of each of the following electromagnetic waves:
a) x-rays
b) microwaves
c) ultraviolet waves.

You need to know the order of the wavelengths of the different types of radiation in the electromagnetic spectrum.

THE BARE BONES

➤ Infrared is used to communicate through optical fibres.
➤ Microwaves enable communication over 'line of sight' distances.
➤ Radio waves can be used to communicate right round the Earth.
➤ Diffraction causes waves to spread out.

A Optical fibre communication

1 Modern optical fibres are so fine that infrared passes straight down the cable.

2 Telephone companies use cables of optical fibres to transmit telephone calls, computer signals and television programmes.

3 An optical fibre carries far more messages than a similar thickness of copper cable.

4 Infrared signals in optical fibres cannot be 'tapped' and suffer less from interference than an electrical signal in a copper cable.

5 Optical fibre cables are important to the 'information revolution' as they carry far more information than equivalent copper cables and are cheaper to manufacture.

Q What are the advantages of optical fibres carrying telephone calls?

B Microwave communications

1 Microwaves are used to transmit calls from mobile phones.

2 Microwaves are used to transmit telephone calls across the country from transmitter to transmitter.

There is some uncertainty about the safety of signals – do they warm up the brain or change brain cells in some way?

3 The microwaves travel in straight lines and the curvature of the Earth limits the distance between transmitters to about 40 km.

4 Satellites are used as relay stations to send signals around the world. Microwaves carry the signals to an orbiting satellite, which receives the signal and transmits it on its way.

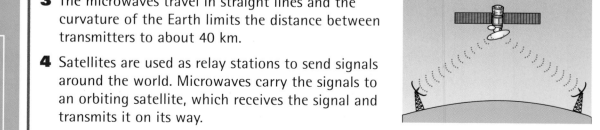

Q Why does there need to be a microwave transmitter every 40 km?

C Diffraction of waves

1 When waves pass through a gap in a barrier, they spread out. This is called **diffraction**.

2 The most diffraction occurs when the wavelengths are the same size as the gap.

3 Diffraction gives us more evidence that light, water waves and sound all behave in the same way.

• We can hear what is being said in another room, even if the people talking are out of sight. Sound is diffracted as it passes through the doorway.

4 Diffraction is important for **television and radio transmissions** to diffract past buildings, mountains and hills to reach houses in more remote locations.

When does iffraction appen best?

D Digital signals

1 A conventional telephone converts the sound waves to a changing electrical signal whose pattern matches the original wave. This is called an **analogue signal**.

2 Modern telephone systems convert the analogue electrical signal to a digital signal – a stream of numbers that describes how the analogue signal changes.

3 Optical fibres or radio waves carry lots of information, television channels or telephone conversations using digital signals.

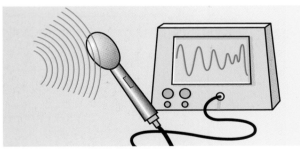

What are e advantages ' using digital gnals?

• Digital signals (including emails and faxes) are damaged less by interference.

4 An analogue signal can have any value. A digital signal can only have certain values.

RACTICE

1 What are the advantages of using infrared through optical fibres for communications?

2 Describe what you would see on a screen as light passed through a very narrow slit.

3 Describe the difference between analogue and digital signals.

Sound and ultrasound

THE BARE BONES

➤ Sound is carried by the particles vibrating in a medium.

➤ Sound cannot travel through empty space.

➤ The pitch of a musical note depends on the frequency of vibration.

➤ Ultrasound is sound at a pitch too high for us to hear.

A   Making sounds

KEY FACT

1   Sounds are made when objects vibrate.

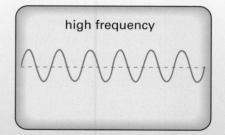

- When a ruler vibrates, the air in front of the ruler is made to vibrate. The sound spreads through the air.

- The particles vibrate with the same **frequency** as the sound and in the same direction as the sound travels.

- The energy of the sound causes other objects to vibrate. The sound makes our eardrum vibrate so we detect sounds.

2 An oscilloscope (CRO) shows the frequency and the **amplitude** of the vibrations:

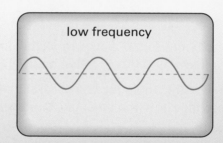

low frequency

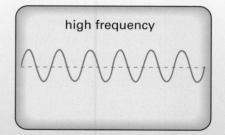

high frequency

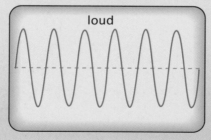

loud

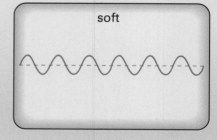

soft

- The lower the frequency of the sound, the lower the pitch of the note.

- The bigger the amplitude of the vibration, the louder the sound. If the sound is too loud, the large vibrations may damage our ears.

3 Example Tightening a guitar string increases the frequency of the vibrations. As the frequency gets higher, the pitch of the notes also gets higher.

Remember
The frequency of the sound is the number of complete vibrations per second. Frequency is measured in hertz (Hz).

Q What kind of wave is a sound wave?

If you can't remember the relationship between frequency and pitch, think of twanging a rubber band. A thick, loose band vibrates slowly and gives a low note. Tighten it and the vibrations get faster, and the pitch higher.

B Echoes and ultrasound

1 An echo is a sound wave being reflected from a hard surface.

2 Bats produce a very high-frequency sound – above 20 000 hertz. This is too high for us to hear, so we call these high-frequency sounds, ultrasound.

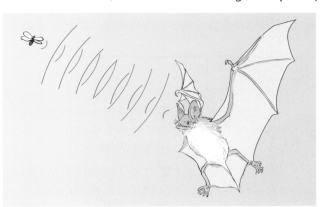

Bats use echolocation to detect objects around them. The bat makes 'clicks' that echo off a moth.

Example The time delay between click and echo is 0.1 s.

Sound travels at 330 m/s. The sound travelled 33m in 0.1 s (there and back). The moth is 16.5 m away.

Dolphins and submarines use echolocation too.

3 Ultrasound is used to 'look' inside the body. Ultrasound is partly reflected when it meets the boundary between two different organs.

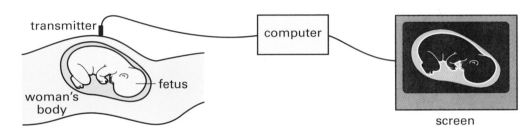

The reflected wave is used to build up a picture of the inside of the body. Ultrasound scans are used to look at babies before birth and also to look at other organs.

4 Ultrasound vibrations can be used for cleaning too. Small objects such as jewellery and electronic components are placed in a cleaning fluid. Ultrasonic vibrations pass through the fluid, causing the dirt particles to shake loose.

5 **Example** A diving vessel uses echolocation to detect submerged objects. An ultrasound beam is transmitted to the bottom of the seabed. It is reflected from a wreck 450 m below. The reflection is detected 0.6 seconds later. How fast does sound travel in sea water?

$$speed = \frac{distance\ travelled}{time\ taken} = \frac{2 \times 450\ m}{0.6\ s} = 1500\ m/s$$

PRACTICE

1 a) A ruler makes 100 vibrations in two seconds. What is the vibration's frequency?
   b) How could you use the ruler to make a higher-pitched note?

2 A bat sends a signal out. It is reflected from the wall of the cave 0.2 seconds later. The cave wall is 30 m away. What is the speed of sound?

3 What evidence is there that sound waves are diffracted?

Structure of the Earth

THE BARE BONES

➤ The thin outer layer of the Earth is broken into large plates that move over the surface.

➤ The rock record provides evidence for the movement of the plates.

➤ Waves caused by earthquakes give evidence of the Earth's structure.

A Structure of the Earth

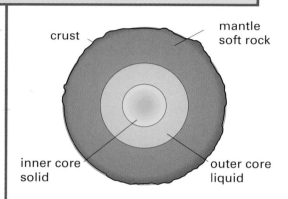

crust
mantle soft rock
inner core solid
outer core liquid

1 The Earth has a layered structure. The thin outer part, the **lithosphere** is about 70 km thick.

2 The lithosphere is broken into large **tectonic plates** that move by slow **convection currents** deep within the Earth.

3 The plates can move in three different ways, each has different effects on the Earth.

- Where the plates move apart, hot molten rock rises and solidifies to form new plate material.

- Where the plates move together, mountains may form by folding, volcanoes may erupt and earthquakes occur.

- Where plates slide past each other, earthquakes occur.

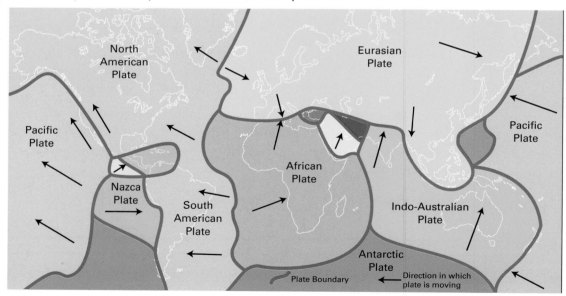

North American Plate

Eurasian Plate

Pacific Plate

Pacific Plate

African Plate

Nazca Plate

South American Plate

Indo-Australian Plate

Antarctic Plate

Plate Boundary — Direction in which plate is moving

4 Example How does the movement of the tectonic plates create new rock?

As the plates move apart, molten rock from deep in the Earth rises to the surface and solidifies.

Q What are the three different ways in which plate movement can affect the Earth?

B The rock record

The rock record provides evidence to support the ideas of plate tectonics.

1 Africa and South America have similar coastlines. Fossil remains of similar plants and animals are found on both continents. It is unlikely that they could all have crossed the Atlantic Ocean, so it is assumed that they were very close at some point in history.

2 New plate material forms under the sea along the Atlantic Ridge. As this rock solidifies, it records the direction of the Earth's magnetic field. The direction of this field 'flips' periodically. The magnetic patterns give information about the way the plates were formed.

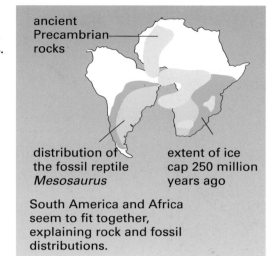

ancient Precambrian rocks

distribution of the fossil reptile *Mesosaurus*

extent of ice cap 250 million years ago

South America and Africa seem to fit together, explaining rock and fossil distributions.

> **Where are the plates moving apart to form the Atlantic ridge?**

C Waves through the Earth

1 Earthquakes occur when the moving tectonic plates stick and then move suddenly. This happens at the earthquake epicentre.

2 The shock from large earthquakes shakes the ground surface and can destroy buildings. Some of the energy from the earthquake is carried through the Earth by seismic waves.

3 Seismometers are used to detect the waves.

4 Seismologists use information about the types of waves that are transmitted and how long they take to arrive at different 'listening stations' to learn more about the structure of the Earth.

- Seismic primary (P) waves are longitudinal waves. They travel quickly and arrive first at the listening stations.

- Seismic secondary (S) waves are slower, transverse waves.

5 Example The energy of the wave when it reaches the Earth's surface causes volcanoes to erupt, the ground to move and may damage buildings.

Check with your teacher how much you need to know about tectonics and earthquakes.

> **What are the two main types of seismic wave?**

PRACTICE

1 Write down two differences between P waves and S waves.

2 An earthquake in Turkey is detected 3000 km away. Which waves, S or P will arrive first?

The solar system

THE BARE BONES

➤ The solar system consists of the Sun, nine planets, the asteroid belt and a number of comets.

➤ All the bodies in the solar system are held in orbit by gravity.

A   The solar system

KEY FACT

The solar system consists of the Sun, nine planets, the asteroid belt and a number of comets.

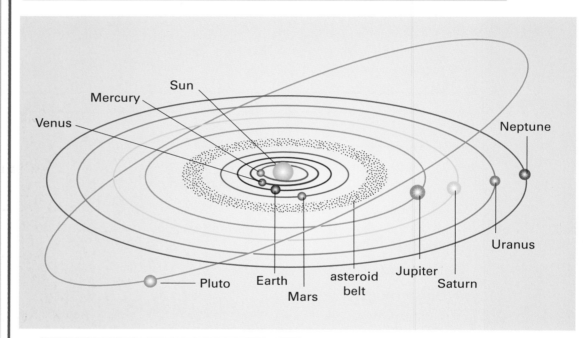

Sun
Mercury
Venus
Neptune
Pluto
Earth
Mars
asteroid belt
Jupiter
Saturn
Uranus

Remember
You should be able to recall the names of the planets.

KEY FACT

1   <u>Nine</u> planets orbit the Sun.

2 The orbits of all the planets are slightly squashed circles (ellipses).

3 The orbit of Pluto is more squashed and at an angle to the orbits of the other planets.

4 The planets are different sizes.

5 The planets take different times to orbit the Sun.

6 Between the four inner planets and the outer planets lies the asteroid belt, made up of dust and rocks. Other asteroids orbit further from the Sun, and some come much closer to the Earth.

Q List the planets in order of distance from the Sun.

Devise a way of recalling the order of the planets from the Sun e.g. My Very Easy Method Just Speeds Up Naming Planets. (Now you have to match the planet names to the first letter of each word!)

A

planet	number of natural satellites	diameter in kilometres	average distance from the Sun in millions of kilometres	time to orbit Sun in Earth units
Mercury	0	4880	58	88 days
Venus	0	12100	108	225 days
Earth	1	12800	150	365 days
Mars	2	6800	228	687 days
Jupiter	16	143000	778	11.9 years
Saturn	18	121000	1430	26.5 years
Uranus	15	52000	2870	84.0 years
Neptune	8	49400	4500	165 years
Pluto	1	3000	5900	248 years

EY FACT

1   Comets are lumps of ice, dust and gas.

2 The orbits of comets are much less circular. They pass close to the Sun, when they can be seen, and then travel far beyond Pluto.

3 When comets are close to the Sun they move faster, when they are far away they move more slowly.

4 When comets pass close to the Sun, some of the ice melts, releasing a tail of dust and gas, which glows in the heat of the Sun.

The Earth and some other planets have one or more moons, which are natural satellites, held in orbit by gravity. Our Moon makes one orbit of the Earth every month.

Example Suggest two reasons why the comets are only visible when they come close to the Sun.

When the comets come close to the Sun, they are closer to the Earth too, so are easier to see. When the comets are a long way from the Sun, they are only seen by reflected light. When they come close to the Sun the ice melts and the gases and dust become so hot they glow.

How many anets have oons?

PRACTICE

1 The orbit of the asteroid Chiron is about 2050 million kilometres from the Sun. Where does that place it in the solar system?

2 State a relationship between the length of time a planet takes to orbit the Sun and the distance the planet is from the Sun.

THE BARE BONES
> All the bodies in the solar system are held in orbit by gravity.
> Artificial satellites orbit the Earth.

A Gravity

1 All masses are attracted to all other masses by the force of gravity.

- The force of gravity is only noticeable if one of the masses is very big – such as the Earth or Moon.

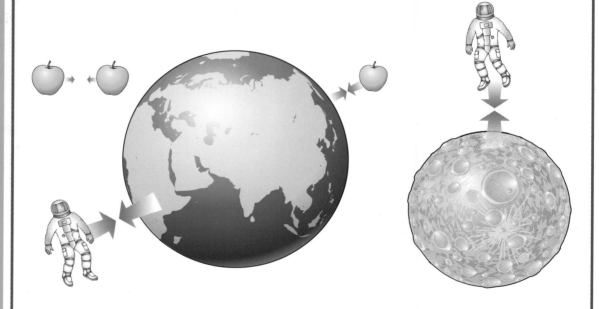

2 All the bodies orbiting the Sun are held in orbit by the force of gravity.

3 The time taken for a planet to orbit the Sun is its year.

- The further a planet is from the Sun, the weaker the pull of gravity on it and the greater its orbit.

- The further from the Sun a planet is, the longer its year.

4 Example Explain why planets further from the Sun take longer to orbit the Sun.

Answer The distance round the orbit is greater, so the orbit takes longer. The pull of gravity is weaker for the distant planets, so the planet moves more slowly.

Q Why is gravity less on the Moon than on the Earth?

B Artificial satellites

1 Artificial satellites are put into orbit around the Earth. They are held in orbit by gravity.

• Satellites close to the Earth orbit much more quickly than those in higher orbits.

• To stay in orbit at a particular distance the satellite must move at a particular speed.

2 Communications satellites are put into a high orbit around the Earth.

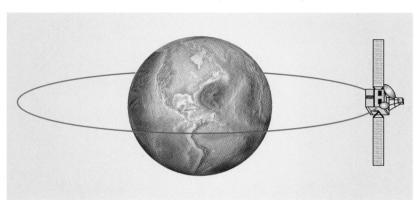

At 36000 km above the Earth they orbit around the Earth at exactly the same rate at which the Earth turns. So they appear from the Earth to remain in one place. This is a **geostationary** orbit.

3 Monitoring satellites study what is happening on the Earth.

• Satellites in orbit over the North and South Poles scan the Earth each day.

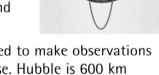

• Monitoring satellites are closer to the Earth, giving better images of the Earth.

• Monitoring satellites are used to observe the weather, land use and the movement of armies.

4 Other satellites, such as the Hubble telescope, are designed to make observations of the rest of the Solar System and deep into the Universe. Hubble is 600 km above the Earth.

5 Example Suggest why it is necessary for communication satellites to remain in the same position above the Earth.

The satellite will remain in contact with the same transmitters and receivers on the Earth 24 hours a day.

List three different uses of satellites.

PRACTICE

1 The asteroid Eros is about 33 kilometres long, 13 kilometres wide and 13 kilometres thick. It is said that a basketball player with a 1 metre vertical leap on Earth could jump nearly 2 kilometres high on Eros! Explain why this might happen.

THE BARE BONES

➤ The Sun is a star in the Milky Way Galaxy.
➤ The Milky Way is one of many millions of galaxies in the Universe.
➤ Stars are formed when clouds of dust and gas are pulled together by the force of gravity.

A Life cycle of a star

KEY FACT

Stars are formed from massive clouds of dust and gas in space.

- Gravity pulls the dust and gas together. As the mass falls together it gets hot. A star is formed when it is hot enough for the hydrogen nuclei to fuse together to make helium. The fusion process releases energy, which keeps the core of the star hot.

- During this stable phase in the life of the star, the force of gravity holding the star together is balanced by the high pressures caused by the high temperatures. Our Sun is at this stable phase in its life.

- When all the hydrogen has been used up in the fusion process, larger nuclei begin to form and the star may expand to become a red giant.

- When all the nuclear reactions are over, a small star, like our Sun, may begin to contract under the pull of gravity. It becomes a white dwarf, which fades and changes colour as it cools down.

- A larger, more massive, star may go on making nuclear reactions, getting hotter and expanding until it explodes as a supernova. The supernova throws dust and gases into space, leaving a small, dense neutron star, which shrinks to a black hole.

- When a star reaches the end of its life it may explode into a cloud of dust and gas. Some of this dust and gas will go on to make new stars. Our Sun was probably created from the dust of an old supernova.

Q Draw a flow chart to show the possible life routes of a star.

Example Why might it be said that 'we are all made of stardust'?

Answer We are made from elements originating on the Earth. The Solar System is formed from the dust of old stars – so we really are stardust.

You might be asked to evaluate evidence for the possibility of life elsewhere in the Universe.

B Beyond the Milky Way

1 The Solar System is centred around our Sun, which is just one of many stars that make up the galaxy called the Milky Way (see picture below). We can see the stars that are fairly close to us as individual points of light. The Milky Way appears as a faint band of light in the night sky. The stars in our galaxy are millions of times further apart than the planets in the Solar System.

2 Our galaxy is just one of many millions of galaxies in the Universe. Some of the galaxies appear as points of light in the night sky. The galaxies are often millions of times further apart than the stars within a galaxy.

3 Observations of light from other galaxies show that their light appears to be shifted towards the red end of the spectrum. The further away a galaxy is, the greater the **red shift**.

4 An explanation for these observations is that all the galaxies in the Universe are moving apart very quickly. The distant galaxies, with the bigger red shift, are moving away faster than those nearer to us. This suggests that the Universe was formed by a **big bang**, which threw all the matter out in different directions.

What does the red shift suggest about the Universe?

C Is there life out there?

1 With all these galaxies containing all these stars, is there another star out there with a planet supporting intelligent life?

2 Physicists are using the best telescopes they have to look for possibilities of life beyond Earth. So far they have discovered over 70 planets orbiting stars other than our Sun.

• Physicists are analysing the radio waves coming from all parts of the Universe. Are any of the signals a message from beyond our solar system?

• The atmosphere of a planet supporting life might be very different from that of a dead planet – the Earth's atmosphere has far more oxygen than it would have if there were no living organisms.

What evidence would you want to see to be convinced that there is life on another planet somewhere in the Universe?

PRACTICE

1 What is the likely end to our star, the Sun?

2 Light from the Sun takes a little over eight minutes to reach us. Light travels at 300 000 kilometres/second. How far away is the Sun?

Work, power and energy

THE BARE BONES

➤ When a force makes an object move, energy is transferred.

➤ Power measures how quickly energy is transferred.

➤ Gravitational potential energy is gained when an object is lifted up.

➤ Kinetic energy is the energy an object has because it is moving.

A Work

1 When you lift a load you do some work – if you lift a heavy load your arms will remind you that you are doing a lot of work. You are transferring energy from your muscles to the weight.

2 Units of work and energy are joules (J) and kilojoules (kJ). 1 kilojoule = 1000 joules.

KEY FACT

3 Use this equation to calculate the work done when a force moves something through a distance:

work done (joules) = force (newtons) × distance (metres)

4 Example A weightlifter lifts 100 kg by 1.5 metres. How much work does he do?

He needs to exert a force of 1000 N to lift 100 kg.

work done = force × distance

work done = 1000 N × 1.5 m
= 1500 J

Q How much work is done when a force of 500 N is used to lift a sack of potatoes 2 m onto a lorry?

B Power

1 Power measures how quickly energy is transferred.

2 Calculate the power from the equation:

$$\text{power (W)} = \frac{\text{work done or energy transferred (J)}}{\text{time taken (s)}}$$

3 A boy running up stairs needs energy for his body to do the work. That energy came from food. But not all the energy from food will have gone in to getting him upstairs. The boy will feel pretty hot too! In fact the human body is only about 25% efficient. There is more about efficiency on page 119.

Q Where does the energy come from to power the boy's muscles?

C Potential energy

1 When the student puts the cans on the shelf he is doing **work**. Energy is transferred from the food in his muscles to the cans.

2 The cans store energy in the Earth's gravitational field. The cans have gained **gravitational potential energy**.

3 The gain in gravitational potential energy depends on the height the cans are lifted and their weight:

$$\begin{array}{ccccc} \text{potential energy} = & \text{mass} \times & \text{gravitational field strength} \times & \text{height lifted} \\ \text{(J)} & \text{(kg)} & \text{(N/kg)} & \text{(m)} \end{array}$$

$$PE = m\,g\,h$$

**How much
otential energy
gained by a
kg person
ho climbs a
eight of 3 m?**

D Kinetic energy

1 What happens to that potential energy when the cans fall? The cans fall to the ground, accelerating as they go.

The moving cans have **kinetic energy**.

2 The kinetic energy depends on the mass of the cans and how fast they fall:

$$\text{kinetic energy (joules)} = \tfrac{1}{2} \times \text{mass (kilograms)} \times \text{speed}^2 \text{ (metres per second)}^2$$

$$KE = \tfrac{1}{2}\,m\,v^2$$

3 **Example** A brick falls from the roof of a building. As it hits the ground it is moving at 20 m/s. The brick has a mass of 4 kg. Calculate the kinetic energy of the brick when it hits the ground.

$$KE = \tfrac{1}{2}\,m\,v^2 = \tfrac{1}{2} \times 4 \text{ kg} \times (20 \text{ m/s})^2 = 800 \text{ J}$$

**What
ppens to the
netic energy
the cans
hen they hit
e ground?**

RACTICE

1

For each picture say whether the object has kinetic energy and/or gravitational potential energy. Explain where the energy came from to get the object to where it is now.

Energy transfers

THE BARE BONES

➤ Energy is transferred when there is a temperature difference between two bodies, i.e. when one is hotter than the other.

➤ Conduction transfers energy from a hot part of a solid to a cold part.

➤ Convection transfers energy by the movement of a liquid or gas.

A Conduction

KEY FACT

1 Energy is transferred from the hotter part of a solid to the colder part by <u>conduction</u>.

Q Identify the good conductors and the good insulators in the pictures in section A.

2
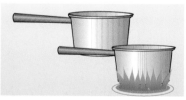

Energy is transferred by particles in the material. The particles in the hot part are vibrating more. These vibrations are passed on to the cooler particles next to them, so the energy spreads through the material until all particles have the same energy.

3 Insulators keep things cold as well as hot.

4 Several layers of clothing will keep you warmer than one thick layer because the layers trap air between them. The air is a good insulator.

No energy transfer is 100% efficient – energy is often lost in warming up the surroundings.

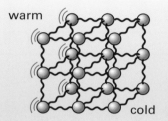

• Metals are good **conductors**, whereas most non-metals are poor conductors. Poor conductors are used as **insulators**.

• Most liquids and gases are poor conductors.

B Convection

1 Convection transfers energy by the movement of a liquid or gas.

Q Explain why the upper floors in a house are often warmer than downstairs.

2 When a liquid or gas becomes warm it expands and becomes less dense. The warmer fluid floats above the cooler fluid, which sinks. This creates a flow, which is called a **convection current**.

3 Most rooms are heated by convection currents. Warm air rises from a heater and cooler air replaces it and is heated.

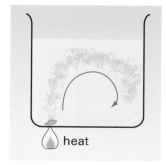

heat

C Evaporation

1 Evaporation is when the particles near the surface of a liquid leave the liquid and become a vapour.

- The escaping particles take some energy with them, so leaving the liquid cooler.

2 You can feel this cooling effect if you put a few drops of ethanol or perfume on your skin. The ethanol evaporates, taking some of the warmth from your skin with it.

3 Example How does sweating keep you cool?

When you sweat your body is using cooling by evaporation. The evaporating sweat carries away energy leaving you cooler.

How does vaporation ause cooling?

D Radiation

All bodies emit radiation. The hotter the body, the more energy it radiates. The radiation is the infrared part of the electromagnetic spectrum.

Y FACT

emember ergy is ansferred from ything that is tter than its rroundings.

- Dark, dull surfaces emit more radiation than light shiny surfaces. They also absorb radiation well.

- Radiation can pass through space – that is how the warmth of the Sun reaches the Earth.

- Light, shiny surfaces do not absorb radiation well – they are good reflectors.

Example Why will it be much hotter inside a dark car on a hot sunny day than inside a white car?

The dark car will absorb more infrared radiation from the Sun, making it hot inside.

Can you uggest ethods of oking that se each of: nduction, onvection and diation?

RACTICE

1 Suggest materials that could be used to insulate the roof space in a house.

2 Suggest why the central heating radiators in houses should perhaps be called 'convectors'.

3 Describe three ways in which energy is lost from the house and suggest ways of reducing the losses.

Energy resources

➤ A lot of the energy we use comes from fossil fuels.

➤ Fossil fuels are non-renewable resources.

➤ Renewable sources of energy are those that are continually replaced, such as the Sun, wind and waves.

A Generation and transmission of electricity

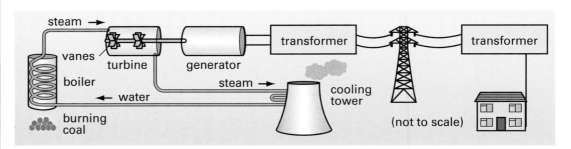

steam → vanes turbine generator boiler burning coal ← water steam → cooling tower transformer transformer (not to scale)

1 Electricity is a very convenient way of supplying energy.

2 Most of our power stations use fossil fuels – coal, oil and gas.

• Fuel is burnt and the energy is used to boil water, to produce steam.

• The steam is used to drive turbines, which turn generators.

• The generators produce electricity.

3 Electricity is transmitted around the country on the National Grid – a network of high-voltage cables. Transformers at the power station step up the a.c. electricity to a high voltage for transmission on the National Grid.

4 As well as producing electricity, fossil fuel power stations produce pollution, excess carbon dioxide and warm up the surroundings.

5 Nuclear power stations do not produce polluting gases. Very little radiation escapes to the surroundings when the power station is running normally. However, the waste from nuclear power stations remains radioactive for many years.

> **Q** We also use fossil fuels to power our cars. What has happened to the energy from the burning petrol by the end of the journey?

B Generating electricity – the future

1 Energy resources that do not get used up or that are continually being replaced are called <u>renewable energy resources</u>.

2 Wave power, wind power and solar panels are renewable energy sources, but they depend on the weather, so cannot be relied on alone. Hydroelectric power stations store energy in a reservoir until demand for electricity is high. The water is then released and electricity generated.

> **Q** Why should we try to reduce our use of electricity?

c The green house

We could make more efficient use of energy and materials in our homes.

Remember
You need to be able to carry out calculations about efficiency.

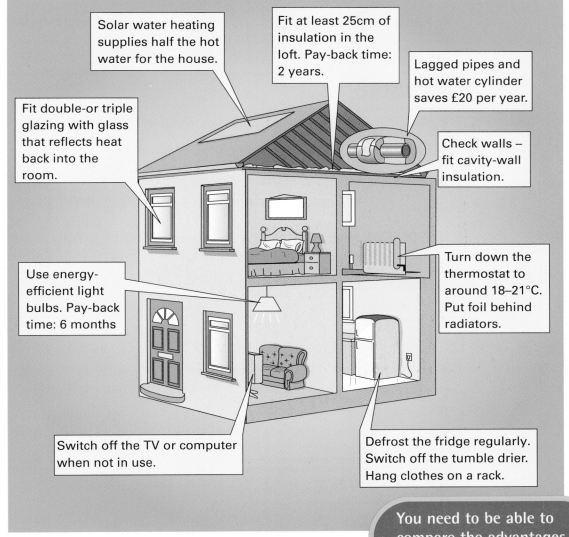

Solar water heating supplies half the hot water for the house.

Fit at least 25cm of insulation in the loft. Pay-back time: 2 years.

Lagged pipes and hot water cylinder saves £20 per year.

Fit double-or triple glazing with glass that reflects heat back into the room.

Check walls – fit cavity-wall insulation.

Use energy-efficient light bulbs. Pay-back time: 6 months

Turn down the thermostat to around 18–21°C. Put foil behind radiators.

Switch off the TV or computer when not in use.

Defrost the fridge regularly. Switch off the tumble drier. Hang clothes on a rack.

Example Identify three places in the home where reducing heat losses can save money.

Double glazing; lagging hot water pipes and hot water tanks; loft insulation; cavity wall insulation.

You need to be able to compare the advantages and disadvantages of using different types of energy sources.

How does turning down the thermostat save energy?

PRACTICE

1 What else do we use fossil fuels for, apart from burning as a fuel?

2 A coal-fired power station has an efficiency of 35%. What does this mean?

3 We could use public transport more and private cars less. How would using public transport save energy?

4 What are the advantages and disadvantages of using wind or water to generate power?

What is radioactivity?

THE BARE BONES

➤ Radioactivity is a random process that takes place in the nuclei of some elements.

➤ Radiation ionises molecules in the material it passes through.

A Radioisotopes

1 Nuclei contain **protons** (positively-charged) and **neutrons** (neutral).

• Each element has a different number of protons.

KEY FACT

2 Some nuclei are unstable because they have <u>too many neutrons</u>. These isotopes are radioactive and are called <u>radioisotopes</u>.

Remember
Mass number
A = number of protons + number of neutrons.

3 Isotopes of an element have the same number of protons, but each isotope has a different numbers of neutrons, so the mass number will be different.

Hydrogen has three isotopes:

hydrogen   deuterium   tritium

Q What are the two types of particles that make up the nucleus of every atom?

4 Symbols are used to represent isotopes. The chemical symbol for the element is used, together with the mass number and the atomic number.

5 Carbon has several isotopes. Carbon-12 is the most abundant isotope. Carbon-14 is a radioisotope created when atmospheric carbon-12 is bombarded by cosmic rays.

B Alpha radiation α

KEY FACT

1 An <u>alpha particle</u> is the same as a nucleus of helium.

• Alpha particles have two protons (and two positive charges) and two neutrons.

2 When an alpha particle is emitted from a radioisotope the nucleus loses two protons and two neutrons – its atomic number decreases by two, its mass number decreases by four. It has become a different element. This is called **transmutation**.

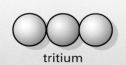

radium – 224   radon –220   +   helium –4
                                                                    alpha particle

3 Alpha particles ionise the air as they collide with the air molecules.

• Each time they ionise an air molecule, alpha particles lose kinetic energy. They can only travel a few centimetres before losing all their kinetic energy. Alpha particles are stopped by paper.

Q What is an alpha particle made of?

C Beta radiation β

1 <u>Beta particles</u> are the same as electrons – but from the nucleus.

2 Inside the nucleus of a radioisotope, a neutron decays to form a proton and an electron. A very small particle called an antineutrino is also produced.

neutron   proton   +   electron   +   antineutrino

3 When a radioisotope emits a beta particle, the mass number does not change, but the atomic number increases by one. It has become a different element.

4 Beta particles are smaller and lighter than alpha particles and move more quickly. They ionise air less well, and can travel a few metres before being stopped. Beta particles can pass through paper but are stopped by a few millimetres thickness of aluminium.

5 Example When a beta particle is emitted the number of protons increases by one, as a neutron loses a beta particle to become a proton.

What is the
charge on a
beta particle?

D Gamma radiation γ

1 <u>Gamma radiation</u> is from the very short wavelength part of the electromagnetic spectrum.

2 Gamma radiation is emitted from an unstable nucleus when it loses energy – often after it has recently emitted an alpha or beta particle.

3 Gamma radiation moves at the speed of light and is less ionising than alpha or beta radiation. It has very high energy and is only stopped by very thick lead or even thicker concrete.

4 When a gamma ray is emitted, the nucleus becomes slightly more stable as it has lost some energy.

What stops
gamma
radiation?

PRACTICE

1 Compete the table summarising the properties of alpha, beta and gamma radiations.

radiation	what is it?	how far can it travel in air?	what stops it?
alpha α			
beta β			
gamma γ			

2 A radioactive material emits particles that can pass through paper, but are stopped by a sheet of aluminium. What are the particles?

3 Place the three types of radiation in order of speed, starting with the slowest.

Radioactive decay

THE BARE BONES

➤ Background radiation comes from natural sources and man-made sources in the environment.

➤ Radioactive decay is a random process.

➤ The half-life is the time taken for half the nuclei present to decay.

A   Background radiation

KEY FACT

1   Background radiation is the radiation that is all around us. Most of it is natural, some is man-made.

• It comes from the Sun and outer space – called cosmic rays. People who travel regularly in high-flying aeroplanes are exposed to more cosmic radiation than people on the ground.

• Some rocks, such as granite, emit radiation. Some parts of the country, such as Cornwall and parts of Derbyshire, have radioactive rocks and buildings made from the rocks.

• Some food is radioactive, particularly shellfish that accumulate potassium in their bodies.

• Radiation still lingers in the atmosphere and in the ground from earlier nuclear events – including the bomb dropped at Hiroshima in 1945 and the accident at Chernobyl in 1986.

• Some radiation in the environment is due to waste from hospitals and nuclear power stations.

2 Our bodies are used to these low doses of background radiation, but if the amount of radioactive material in our surroundings becomes too high, there could be problems for our health.

Q How could you reduce your exposure to background radiation?

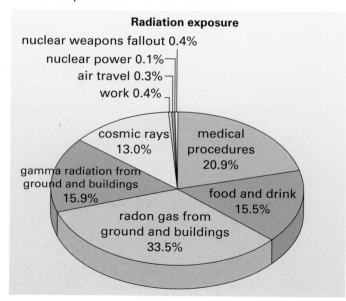

Radiation exposure

nuclear weapons fallout 0.4%
nuclear power 0.1%
air travel 0.3%
work 0.4%

cosmic rays 13.0%
medical procedures 20.9%
gamma radiation from ground and buildings 15.9%
food and drink 15.5%
radon gas from ground and buildings 33.5%

3 Background radiation must be taken into account when taking radioactivity measurements.

4 Example Look at the pie chart. What proportion of the background radiation is due to natural sources?

Answer
15.5% + 33.5% + 15.9% + 13.0% = 77.9%

B Half-life

- Radioactive decay is a random process that helps to stabilise the nucleus.

- You cannot predict when a particular nucleus will decay.

- You can say that after a certain amount of time, half the nuclei in a sample of the material will have decayed.

> The time it takes for half the nuclei in a sample to decay is called the <u>half-life</u>.

- As the nuclei decay, there will be fewer left to decay, so the decay rate will decrease. A graph showing how much radioactive material is present will show the rate at which the material decays.

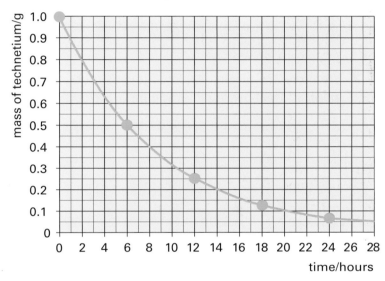

The graph shows how a sample of technetium decays. Technetium-99 has a half-life of six hours. This means that if there is one gram at the start, then after six hours this will have decayed to half a gram and another six hours later there will only be one quarter of a gram left.

- The half-lives of radioisotopes vary from fractions of seconds to millions of years.

- **Example** Look at the graph for technetium. How much is left after 18 hours? How much after 24 hours?

  After 12 hours there was $\frac{1}{4}$ gram, so after another six hours there will be half that – $\frac{1}{8}$ gram. After 24 hours (four half-lives) there will be $\frac{1}{16}$ gram left.

What does half-life mean?

1 A particular radioactive isotope has a half-life of two days. Twelve milligrammes of the isotope are injected into the patient. After six days:
   (a) how many half-lives have passed?
   (b) how much of the isotope will remain in the patient?

2 The half-life of radon-220 is 55 s. How long will it take for sample of radon-220 to lose three-quarters of its radioactive atoms?

THE BARE BONES

➤ The predictable way that radioisotopes decay is used to calculate the age of objects.

A Dating using radioisotopes

1 Radioisotopes decay to make stable isotopes. Physicists can measure how much radioisotope is present in a sample of material and how much of the stable isotope is present. From this they can work out the age of the sample of material.

2 Uranium-235 occurs in some igneous rocks. It gradually decays to lead-207. The graph shows how much uranium-235 would be left from a sample of 100 g.

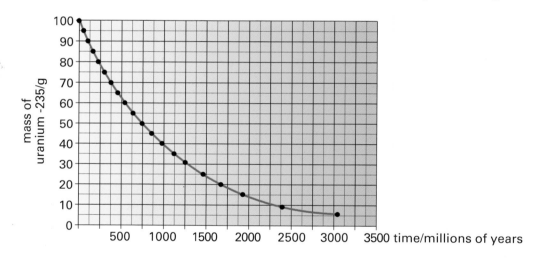

Scientists compare the amount of uranium to the amount of lead in a rock to estimate the age of the rock.

After 1000 million years there would be about 35g of uranium-235 and 65 g of lead-207.

Example Look at the graph for uranium-235. How old would the rock be if there were 60 g of uranium to 40 g lead?

Answer 500 million years-old

A

3 Carbon-14 is created from carbon-12 when cosmic rays bombard the atmosphere. Growing plants take in carbon-14 together with carbon-12 during photosynthesis. While the plant is alive the proportion of carbon-14 is constant. When the plant dies the carbon-14 decays with a half-life of 5730 years.

> Using these facts to find out how old the plant material is, is called radio-carbon dating.

4 In Turin, Italy is a piece of cloth that some people believe was used to wrap the body of Jesus Christ after he died (see below). Could it really have belonged to Christ? In 1988 scientists in three different laboratories took very small samples of the cloth to find out how old it was. They used radio-carbon dating techniques because the cloth was linen – made from the flax plant.

They all agreed that the flax from which the cloth was made, was growing some time around the year 1325 AD – long after the death of Christ.

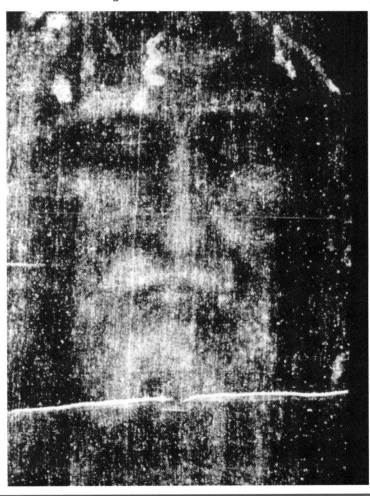

Why would radio-carbon dating be of no use in trying to date something made from metal?

PRACTICE

1 Use the uranium-235 graph to find the half-life of uranium-235

2 Suggest why three different laboratories all tested the Turin shroud at the same time.

Using radioactivity

THE BARE BONES

➤ As radiation passes through material it ionises the molecules.
➤ Radiation can damage cells in the body.
➤ Radioactive materials can be used to trace the flow of fluids.
➤ Radiation is used to measure the thickness of materials.

A Radiation safety

KEY FACT

1 Radiation can cause ionisation of the molecules in living cells. This damages the cells and may cause cancer.

2 Care must be taken when handling radioactive materials. People who work with ionising radiations must be particularly careful – the greater the dose, the greater the risk of damage.

You need to be able to explain why a particular radiation is best for the job.

3 Radiation can be detected using a Geiger-Müller tube.

4 People who work with ionising radiation wear a **film badge**. The badge contains photographic film that is sensitive to ionising radiation. The badge is lined with different materials so that different parts of the badge detect different radiation.

Q What properties of radiation are used in its detection?

- aluminium
- thin plastic
- film
- backing

B Using radiation in medicine

1 Large doses of radiation are used to kill cancer cells in the body. Their use is carefully controlled. There is sometimes some short-term damage to other healthy cells.

2 The different penetrating powers and ionisation properties are taken into account in choosing the radiation for the task.

Q Explain why beta radiation is not used for treating cancers from outside the body.

3 The body would absorb beta radiation emitted from outside the body before it reached most cancers. However, iodine-131, a beta emitter, is used to treat cancer in the thyroid. The iodine is injected into the blood stream and taken up by the thyroid gland.

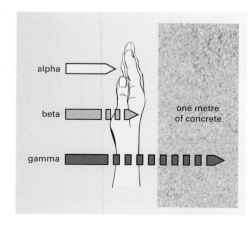

alpha

beta

gamma

one metre of concrete

c Industrial uses of radiation

1 Radiation is absorbed by the material through which it passes. The thicker the material, the greater the absorption. This effect is used to monitor the thickness of materials in production.

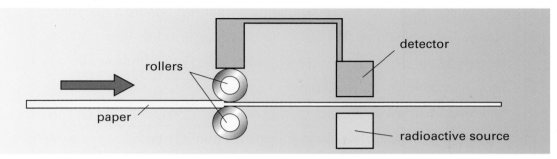

2 Some radioactive materials are used as tracers in industry. The material can be put into a fluid system, such as piped water. The radiation emitted shows where the fluid has flowed. This can be used to identify blockages or leakages.

- A short half-life radioisotope is used so it does not stay in the environment any longer than necessary for the tracing.

- A gamma ray emitter should be used, so that the radiation can be detected outside the pipe.

3 In nuclear power stations, neutrons bombard uranium-235. The nucleus of uranium-235 splits into two smaller nuclei. This is **nuclear fission**.

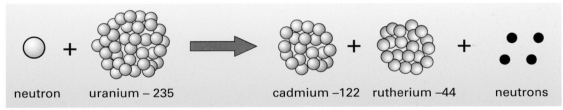

neutron   uranium – 235   cadmium –122   rutherium –44   neutrons

The power station uses the energy released from the nucleus to heat water to generate electricity.

4 Example Suggest, with a reason, which radiation would be best to use to monitor the thickness of sheet metal used to make food cans.

Gamma radiation would be best. The metal would stop some of the gamma radiation, so the reading on a counter would depend on the thickness of the metal. Alpha and beta-radiation would be stopped completely.

Why would gamma radiation be of use to detect paper thickness?

PRACTICE

1 Explain why alpha emitters are not used for the treatment of cancers.

2 Why is a gamma emitter best used for tracing fluid flows?

Biology

1   A group of students went walking in the hills. When they stopped to rest, one boy felt very cold. He began to shiver. His friend suggested that he should curl up into a ball shape.

   a)   Suggest why curling up in a ball helps to reduce heat loss from his body.

   He <u>reduces surface area</u> by curling up his body. **(1)** | These are the key words. |

   b)   How does shivering help to keep his body warm?

   Shivering is caused by <u>muscles moving</u>, and this <u>generates heat</u>. **(2)** | Two points are needed here, usually a <u>fact</u> linked to an <u>explanation</u>. |

   c)   Explain how sweat glands increase heat loss from the body.

   • Sweat contains water

   • Water evaporates.

   • Evaporation takes/requires/transfers heat/energy from the body.

   • Losing heat/energy causes cooling. **(2)**

   | Any two of these points will give you the marks. |

2   Two students, Peter and Kelly, ran an 800 metres race.
Before the race, when they were resting, Peter's pulse rate was 82 beats per minute and Kelly's was 70 beats per minute.
Just after the race, their teacher measured their pulse rates (bpm).
The teacher measured them again at 2 minute intervals.

| Make sure you <u>read all the information</u> because the answers are here. |

The results are shown in the table below:

	Time after race (minutes)				
	0	2	4	6	8
Peter's pulse rate (bpm)	120	100	91	86	83
Kelly's pulse rate (bpm)	100	82	73	70	70

| What do you notice about this set of data compared to Peter's? |

Peter's results are shown in the graph on the opposite page.

   a)   Draw a graph of Kelly's results on the grid. **(3)**

   b)   Recovery rate is one measure of fitness. Explain how the graphs show that Kelly is probably fitter than Peter. **(4)**

   | You need to write at least three different points and put the sentences together properly for an extra mark. |

Kelly's pulse does not rise as high; it has a more rapid fall; <u>reaches expected resting rate more quickly.</u> **(4)**

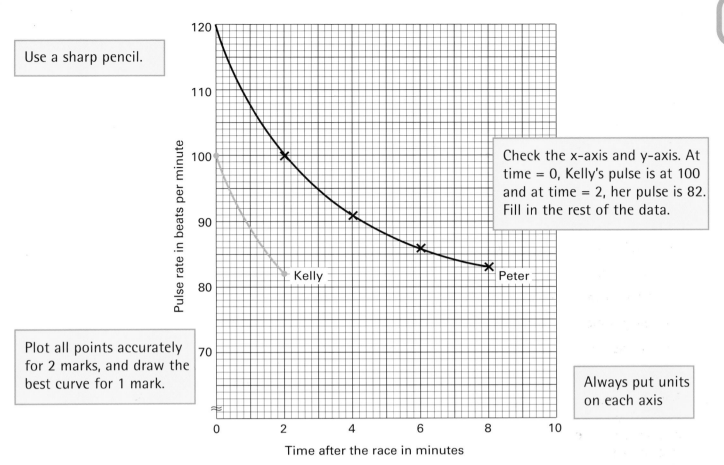

Use a sharp pencil.

Check the x-axis and y-axis. At time = 0, Kelly's pulse is at 100 and at time = 2, her pulse is 82. Fill in the rest of the data.

Plot all points accurately for 2 marks, and draw the best curve for 1 mark.

Always put units on each axis

Further questions to try (answers on page 177).

3 Page 34 shows a diagram of the eye.

List A gives the names of some parts of the eye.

List B gives the functions of these parts in a different order.

Draw a straight line from each part in list A to its function in list B.

One has been done for you.

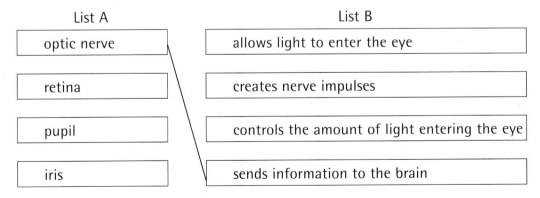

4   A student, called Sarah, was exercising at the gym.

a)   Complete the paragraph using words from the box.

carbon dioxide	nitrogen	oxygen

When exercising, Sarah breathes faster because her muscles use more_____. The air she

breathes out contains more_____than the air she breathes in. **(2)**

b)   The diagram below shows two types of blood vessel.

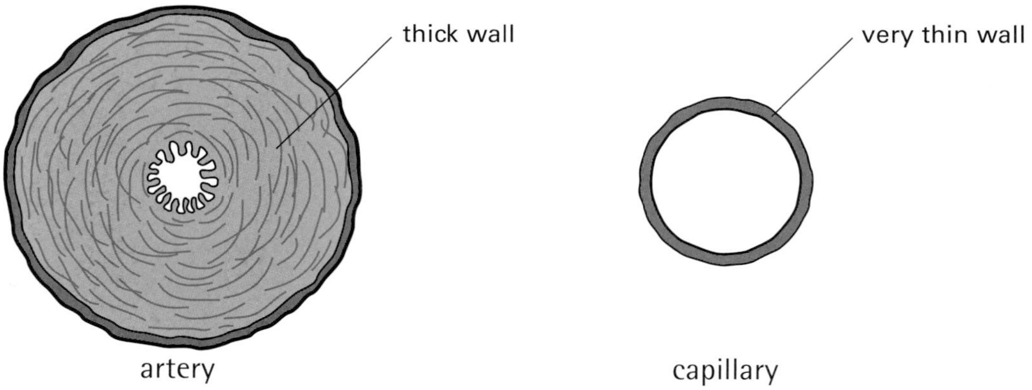

thick wall   very thin wall

artery   capillary

i) How does the thick wall of the artery help with its function?

_____

_____ **(1)**

ii) How does the very thin wall of the capillary help with its function?

_____

_____ **(1)**

c)   A vein is another type of blood vessel. There are valves inside a vein.
      What is the function of these valves?

_____

_____ **(1)**

5   The diagram shows the human digestive system.

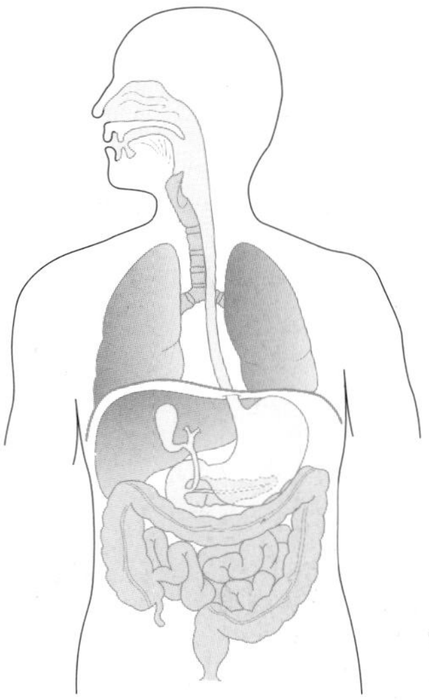

a i) The digestion of starch begins in the mouth.

Name the enzyme that digests starch.

_____ (1)

   ii) In which organ does the digestion of protein begin?

_____ (1)

b)   Digested food is sbsorbed into the blood though the villi. The diagram (below right) shows a section through a villus.

Describe **two** features of villi that help the absorption of digested food.

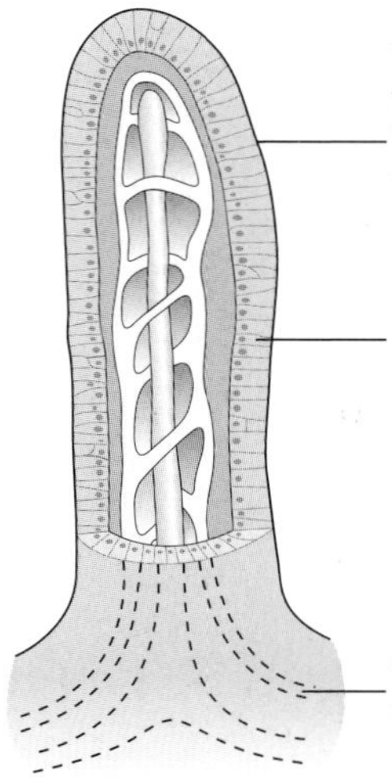

1_____

_____ (1)

2_____

_____ (1)

wall of villus

blood capillaries

vein taking blood to liver

c)   People can get food poisoning from eating food containing harmful bacteria.
One effect of food poisoning is diarrhoea.
This means that only a little water is absorbed from the digestive system into the blood.

   i) On the diagram of the digestive system (above right) shade in the organ where most water is usually absorbed into the blood. (1)

   ii) Explain why body temperature may rise if diarrhoea lasts for a few days.

_____

_____

_____

_____ (2)

6   A plant grows leaves for:

A   making food   B   absorbing water
C   reproduction   D   holding it in the soil.

7 Soil in a wood is covered by decaying leaves. The leaves decay because:

    A birds feed on them   B squirrels make nests from them

    C microorganisms feed on them   D spiders make webs over them.

8 Respiration is the:

    A release of energy from glucose   B pumping of air into the lungs

    C release of sweat onto the skin   D resuscitation of a person who has stopped breathing.

9 The rate of transpiration is affected by the weather.

   Which row of the table shows the effect of it starting to rain and becoming windy?

	effect on rate of transpiration	
	starting to rain	becoming windy
A	decreases	decreases
B	decreases	increases
C	increases	decreases
D	increases	increases

10 The diagram shows parts of a plant cell.

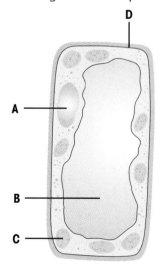

a) Label **two** parts that are found in **both** animal and plant cells.
    Use words from the box.

cell membrane   cell wall   chloroplast   cytoplasm   nucleus

             (4)

b)   Choose the correct word from the box below to complete the following sentence.

cellulose	protein	starch

The cell wall is made of _____ (1)

c)   Stephen took two similar cuttings from the stem of a plant. He dipped the end of one cutting in a plant hormone. Then he planted both cuttings in soil.

After four weeks, he removed both cuttings from the soil.
The diagram shows what they looked like.

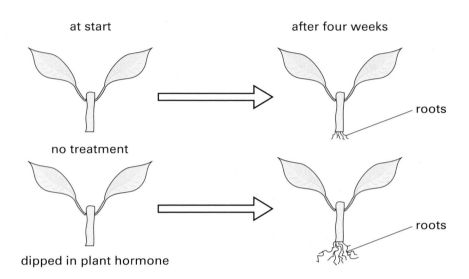

i) What effect did the plant hormone have on the cutting?

_____

_____

_____

_____ (2)

ii) Suggest how this will help the cutting to survive.

_____

_____ (2)

11   Read the newspaper article on the following page and then answer the questions.

Environmentally 'friendly herbicide' found.

Biologists working on The Great Barrier Reef off the coast of Australia have discovered herbicides that are harmless to humans, other animals and crops.

Greenhouse trials show that some substances in reef organisms stop photosynthesis in weeds.

If field trials show the same results as the greenhouse trials, these substances could be a new class of herbicide, which kills weeds without damaging the environment or crops.

The biologist got the idea for their work because they noticed that parts of the reef had no plants. They found about 5000 separate substances in the reef organisms. The biologists tested each substance for its effect until they found the substances they were looking for.

a) Explain how the herbicides from reef organisms kill weeds.

_____

_____ (2)

b) Farmers already use herbicides to kill weeds.
Suggest why herbicides from reef organisms are described as a new class.

_____ (1)

c) Suggest why the first trials were carried out in a greenhouse rather than in a field.

_____

_____ (2)

d) What observations gave the biologists the idea to do the research.

_____ (1)

e) Explain why it was important to test each of the 5000 substances separately.

_____ (1)

Answers to the further questions

3

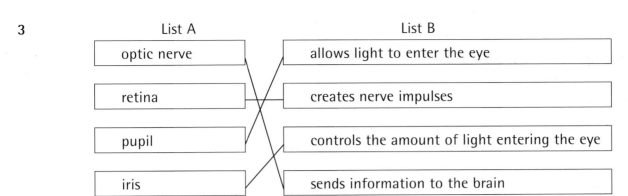

List A	List B
optic nerve	allows light to enter the eye
retina	creates nerve impulses
pupil	controls the amount of light entering the eye
iris	sends information to the brain

4   a)   When exercising, Sarah breathes faster because her muscles use more oxygen. The air she breathes out contains more carbon dioxide than the air she breathes in.

  b i)   Withstands pressure / can pull back into shape / undergoes elastic recoil.

   ii)   Allows diffusion / exchange of materials.

  c i)   The shaded area should be everywhere above the cross section.

   ii)   To prevent the blood flowing backwards.

5   a i)   carbohydrase / amylase   ii)   The stomach

  b)   1. cell walls only one cell thick   2. increased surface area

  c ii)   Any two of: less water in blood / less sweat produced / less evaporation / less heat lost.

6   A

7   C

8   A

9   B

10   a)   Cytoplasm, cell membrane or nucleus.

   b)   The cell wall is made of cellulose.

   c i)   Any two from: longer roots/more roots/larger roots/more spread out.

   c ii)   Either: firmer enchorage, or: better water/mineral/ion absorption.

11   a)   stops photosynthesis/food/sugar production; so cannot respire/produce protein / grow.

   b)   less damage to environment / animals / harmless to humans

   c)   better control over variables/isolated from the environment/so more reliable results

   d)   parts of the reef had no plants

   e)   to establish which one was effective/find the effect of each chemical/analysis with a view to synthesis

1 a) Use words from the list to complete the passage about organic compounds:

> carbon   carbon dioxide   hydrogen energy   fuels   neutral   water   wood

Some organic compounds are used as **fuels** as they release **energy** when they are burned.

All organic compounds contain the element **carbon**.

Many also contain the element **hydrogen**.

When organic compounds containing hydrogen are burned in a plentiful supply of air, the two substances formed are **carbon dioxide** and **water**. **(3)**

   b) Why is it dangerous to burn organic compounds in a limited supply of air?

Carbon monoxide could be formed, which is a poisonous gas. (2)

2 Explain why enzymes are used in industry.

They are used as they bring about reactions at lower temperature and at a lower pressure and the process then becomes less expensive. (3)

> For questions on industrial processes, you need to remember that you want plenty of product being produced cheaply.

3 A small piece of sodium is dropped into a large beaker of water. It reacts to form sodium hydroxide solution and a gas.

Describe **three** things you would **see** in this experiment.

The sodium bubbles, floats in water and burns with a yellow flame. (2)

(Other answers are: moves about/dissolves/gets smaller)

> This is testing your observations, not asking what is produced.

4 Draw the electronic structure of a chlorine atom and a chloride ion.

(Atomic number of chlorine is 17)

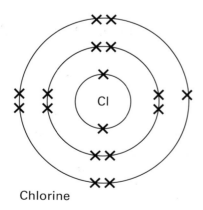

Chlorine

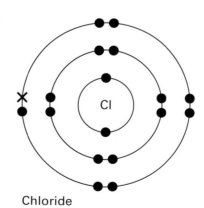

Chloride

> An ion will have more or less electrons than the atom. The ion will always have a full outer shell.

Further questions to try (answers on page 180).

5 Fertilisers are made from ammonia.

a) Why do farmers add fertilisers to their crops?

____ (1)

b) Excess fertilisers are washed off fields into rivers. State two consequences of this.

____

____ (2)

6 Describe how crude oil was formed.

____

____

____ (3)

7 Scientists believe that when the Earth first formed it was very hot, with many volcanoes.

a) Explain how oceans formed on the Earth's surface.

____

____

____ (3)

b) The first green plants on Earth produced oxygen. This caused a gradual increase in the amount of oxygen in the atmosphere. Then animals began to live on the Earth.

Explain why the amount of oxygen in the atmosphere is now constant (does not rise or fall).

____

____ (2)

8 a) Choose words from the list to complete the sentences, below, about magnesium burning in air.

Electrical	heat	light	kinetic	an endothermic
an exothermic		a neutralisation		a reduction

When magnesium burns, it transfers ____________ and ____________ energy to the surroundings.

We say this is ______________ reaction. (3)

b) Complete the word equation: magnesium + __________ → magnesium oxide (1)

Four solutions were tested with universal indicator solution to test their pH.
Complete the following table:

solutions	pH	colour of universal indicator
1	7	
2	4	
3	11	
4	1	

Answers to further questions

5 **a)** To grow bigger or better crops.

b) Increase plant life in the river. Eventually the plants decay and so oxygen is used up in the river.

6 Crude oil was formed by small sea creatures, which sank to the sea bed. After millions of years the effects of temperature and pressure caused crude oil to form.

7 **a)** The volcanoes gave out gases, including steam. The earth cooled and the steam condensed and turned into liquid water to form the oceans.

b) The plants produce oxygen by photosynthesis and animals use the oxygen for respiration. These processes occur at the same rate, creating a balance in gases produced and used up.

8 **a)** When magnesium burns, it transfers heat and light energy to the surroundings. We say this is an exothermic reaction.

b) magnesium + oxygen → magnesium oxide

9

solutions	pH	colour of universal indicator
1	7	green
2	4	orange
3	11	purple
4	1	red

Physics

1 **a)** Write the name of each component under the symbol. Choose your answers from:

LED   resistor   voltmeter

> Make sure you choose words from the list.

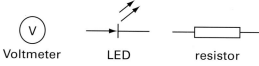

Voltmeter   LED   resistor

b) Anna wires up two lamps and a cell to light her sister's doll's house. Both lamps light normally.

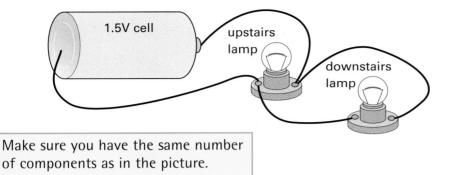

Make sure you have the same number of components as in the picture.

i) Finish the circuit diagram in the space provided, using the correct symbols. Label each component.

ii) Laura wants to switch each lamp on and off separately. She adds two switches to do this.
Finish the new circuit diagram in the space provided. Label each component.

Make sure the switches will work the way the question asks.

circuit diagram

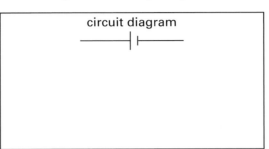

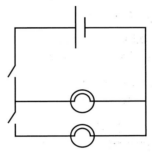

c) Laura used a soldering iron to make the connections. The soldering iron is connected to a 24 V supply. A current of 2 A passes when the iron is switched on.
Calculate the power rating of the soldering iron.

Write down the equation:	$P = V \times I$
Write in the values you know:	$P = 24\ V \times 2\ A$
Write out the answer <u>with the units</u>:	$P = 48\ W$

2 A hovercraft is at rest on land.

a) Draw an arrow on the hovercraft to show where the weight acts.

b) The diagrams below show the forces on the hovercraft at different points in the journey.
The length of each arrow represents the size of the force.

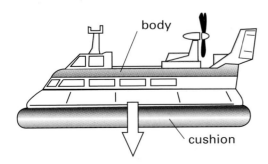

Your arrow should be pointing down and acting from somewhere within the width of the hovercraft.

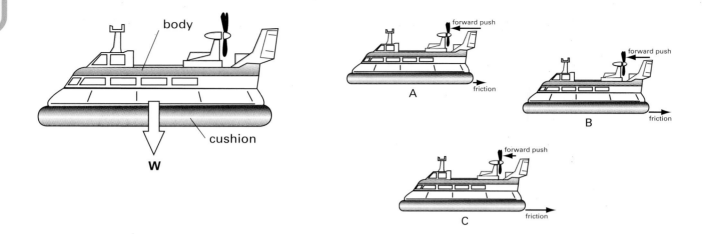

i) Which diagram shows the hovercraft moving at constant speed?

If the hovercraft is moving at a constant speed the forces are balanced – so it must be B.

ii) Which diagram shows the hovercraft accelerating?

If the hovercraft is accelerating the unbalanced force must be forward – so it must be A.

iii) Which diagram shows the hovercraft slowing down?

If the hovercraft is slowing down the friction forces must be bigger than the forward forces, it must be C.

c) The hovercraft travels between Portsmouth and the Isle of Wight. It takes half an hour (0.5 hour) to make the 15 km journey. Calculate the average speed of the journey.

Write down the equation:	speed $= \dfrac{\text{distance}}{\text{time}}$
Write in the values you know:	speed $= \dfrac{15 \text{ km}}{0.5 \text{ h}}$
Write out the answer <u>with the units</u>:	speed = 30 km / h

Further questions to try (answers on page 183)

3 a) Write down the name of the Earth's only natural satellite.

b) What force keeps this satellite in orbit around the Earth?

c) Write down the name of one planet whose orbit is longer than the Earth's.

d) The Earth takes 24 hours to rotate on its axis. A communications satellite appears to stay above the same point on the Earth's equator. How long does the communications satellite take to orbit the Earth?

e) A weather satellite takes 100 minutes to orbit the Earth. How does the height of the weather satellite compare with the height of the communications satellite? Explain your answer.

4 a) A marathon runner has been running for more than 23 miles. He is very hot and sweaty. Sweating helps the runner lose energy. Use your ideas about energy transfer to explain how this happens.

   b) At the end of the race the runner is given a shiny foil blanket. This stops him cooling down too quickly. Use your ideas about energy transfer to explain two ways in which this happens.

5 The inside of a smoke alarm is shown below. It uses alpha radiation to detect smoke.

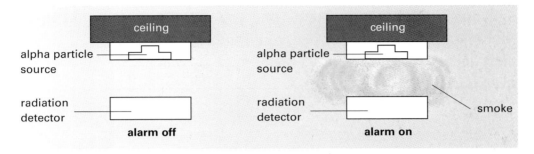

   a) Use your knowledge about alpha particles to explain how it works.

   b) Suggest one reason why a gamma source is not suitable for a smoke detector.

Answers to the further questions

3 a) Moon

   b) gravity

   c) All the planets further from the Sun than Earth take longer to orbit the Earth: Mars, Jupiter, Saturn, Uranus, Neptune or Pluto.

   d) 24 hours – it stays over the same point so must complete an orbit in one day.

   e) The weather satellite completes its orbit more quickly, so must be at a lower height than the communications satellite.

4 a) Sweat evaporates from the surface of the skin. When water evaporates, the vapour molecules have more energy than the liquid molecules on the skin. The vapour molecules carry energy away from the body, leaving it cooler.

   b) The shiny foil blanket is a poor radiator of energy, so he does not lose much energy that way. The blanket prevents draughts carrying away the runner's sweat, so he does not cool down so much.

5 a) The alpha particles can cross the small gap and reach the detector. When smoke is in the gap, the alpha particles are absorbed and can no longer reach the detector and the alarm is set off.

   b) Gamma rays would still reach the detector even through the smoke. They would also reach the people using the room.

Topic checker

- Go through these questions after you've revised a group of topics, putting a tick if you know the answer, a cross if you don't – you can check your answers at the bottom of each page.
- Try these questions again the next time you revise . . . until you've got a column that's all ticks! Then you'll know you can be confident . . .

Biology

Cells and molecules

1 What is growth?	☐	☐	☐
2 All cells have some features in common, but the structure can vary. Why is this useful?	☐	☐	☐
3 What is the main reason for digestion?	☐	☐	☐
4 What are the main types of digestive enzymes?	☐	☐	☐
5 Why are vitamins and minerals important for nutrition?	☐	☐	☐
6 Why is the shape of an enzyme molecule important for its function?	☐	☐	☐

Body systems

7 What are five differences between arteries and veins?	☐	☐	☐
8 Why are valves important in the heart?	☐	☐	☐
9 What are the main components of blood?	☐	☐	☐
10 What is meant by 'immune response'?	☐	☐	☐
11 Why is air pressure important in breathing?	☐	☐	☐
12 Where and when is anaerobic respiration likely to occur in humans?	☐	☐	☐

Answers 1 permanent increase in size; increase in cell number and size 2 cell structure matches function and different cell types are specialised to carry out different functions 3 to reduce particle size of food, so that particles can pass into the bloodstream 4 carbohydrase (breaks down carbohydrates); lipase (breaks down fats/lipids); protease (breaks down proteins) 5 for general health; some are needed for enzymes to work properly 6 active site of an enzyme molecule has a unique shape where particles can 'lock on' and react

7 arteries: more muscular walls, no valves, smaller central canal/lumen, carry oxygenated blood, carry blood away from heart/towards organs (veins opposite to this); speed of blood flow is fast 8 maintain correct direction of blood flow/prevent back flow of blood 9 cells: red and white; platelets/cell fragments; plasma containing dissolved substances and plasma proteins 10 production of antibodies by white cells due to the presence of antigens/foreign proteins 11 difference in pressure inside lungs and outside body causes inhalation/exhalation 12 muscles, during rapid exercise

13 What are the main components of the nervous system? ☐ ☐ ☐

14 How does alcohol affect the nervous system? ☐ ☐ ☐

15 What reflex action does the iris carry out? ☐ ☐ ☐

16 How do insulin and glucagon help to control blood sugar level? ☐ ☐ ☐

17 Which body changes occur at puberty? ☐ ☐ ☐

18 Which hormone causes the production of sex cells in human males? ☐ ☐ ☐

19 How is the structure of the uterus suited to its function? ☐ ☐ ☐

20 What is fertilisation? ☐ ☐ ☐

21 Which body conditions are controlled by homeostasis? ☐ ☐ ☐

22 How does the kidney help in homeostasis? ☐ ☐ ☐

Plants and photosynthesis

23 What are the raw materials and products of photosynthesis? ☐ ☐ ☐

24 Which conditions affect the rate of photosynthesis? ☐ ☐ ☐

25 What is meant by a concentration gradient in osmosis? ☐ ☐ ☐

26 Why is a waxy cuticle found on the upper leaf surface, while stomata are mostly on the lower surface? ☐ ☐ ☐

27 How do auxins cause growth movements (tropisms)? ☐ ☐ ☐

28 How are auxins used commercially? ☐ ☐ ☐

Answers 13 receptors/sensory nerves or sense organs; central processing unit/the brain and spinal cord; motor nerves to effectors 14 sedative, slows response time, affects behaviour and judgement 15 automatically adjusts to amount of light entering eye 16 insulin lowers blood sugar level, mainly by causing the conversion of glucose to glycogen; glucagon does the opposite 17 growth spurt; pubic hair growth; production of sex cells/eggs/sperms; penis and muscle development in males; breast development and widening of hips in females 18 testosterone 19 can expand to a much bigger size, muscular to deliver the baby during birth; protected from surroundings; warm 20 nucleus of sperm and egg fuse 21 water, salt and pH balance in blood and tissue fluids; glucose level of blood; body temperature; waste removal 22 balances water and salt needs of body; produces urine to get rid of excess water and excretes urea 23 raw materials are carbon dioxide and water; products are glucose and oxygen 24 main conditions: light intensity, availability of carbon dioxide, temperature, also minerals can affect the production of chlorophyll, and water (which would be present in living cells) 25 a difference in the concentration of water molecules between one area, or one cell and another 26 upper surface is generally in more direct sunlight so more likely to dry out, and waxy cuticle protects against water loss; stomata allow water loss, so their position on underside of leaf ensures that the rate of loss is minimal 27 increase rate of growth in shoot tips but inhibit growth in root tips; cause uneven growth either side of a shoot or root tip so it curves 28 to stimulate root growth in stem cuttings; regulate fruit ripening; cause abnormal growth and plant death in weedkiller

Variation and inheritance

29 What are the main reasons for variation between individuals?	☐ ☐ ☐
30 What were the main points of Charles Darwin's theory of evolution by natural selection?	☐ ☐ ☐
31 What are the differences between sexual and asexual reproduction?	☐ ☐ ☐
32 How is cloning used in agriculture?	☐ ☐ ☐
33 What is an allele?	☐ ☐ ☐
34 How many alleles for a gene are found in: a) an egg cell, b) a muscle cell of a foetus?	☐ ☐ ☐

Biodiversity and ecosystems

35 What is biodiversity and why is it important?	☐ ☐ ☐
36 Which human activities may have had a bad impact on the: a) atmosphere b) soil c) oceans?	☐ ☐ ☐
37 How are the population sizes of a predator and its prey linked?	☐ ☐ ☐
38 Which processes add carbon dioxide to the atmosphere and which remove it?	☐ ☐ ☐
39 What is meant by 'producers' and 'consumers' in a food web?	☐ ☐ ☐
40 How is a pyramid of biomass drawn?	☐ ☐ ☐

Chemistry

Atomic structure

41 What are the three particles in an atom?	☐ ☐ ☐
42 What are the charges of these particles?	☐ ☐ ☐

Answers 29 genetic, due to inheritance of different DNA/genes/chromosomes; environmental, due to factors in the environment/lifestyle 30 genetic variation gives rise to new characteristics; some characteristics are beneficial for survival; better-adapted individuals are more likely to reproduce and pass on beneficial characteristics 31 sexual: involves two parents, each providing a nucleus from a sex cell; genetic variation in offspring; asexual: involves one parent, offspring genetically identical and identical to parent 32 used to produce many more individuals the same as the original, e.g. splitting embryos in cattle and growing cauliflowers from tissue culture 33 one form of a gene 34 a) one b) two or a pair 35 variety of living things / represents a gene pool genetic variation between living things 36 a) burning fossil fuels; release of chemicals, such as PCBs b) acid rain caused by burning fossil fuels; deforestation; contamination with chemicals c) release of sewage and chemicals 37 dependent on each other; predator population follows trend for prey population, but after a time lag 38 add carbon dioxide: respiration, burning/combustion of carbon-based compounds, decay/rotting; remove carbon dioxide: photosynthesis 39 producers are plants / organisms that carry out photosynthesis; can manufacture own food from simple raw materials; consumers are animals / organisms that need a ready-made food source 40 using a set unit on the horizontal scale for a unit of mass, e.g. 1 cm = 1 kg
41 Electrons, protons and neutrons 42 Protons are positive, electrons are negative, neutrons have no charge.

43 What information does the atomic number give? ☐ ☐ ☐

44 What information does the mass number give? ☐ ☐ ☐

45 How many electrons can fit into each of the first three shells? ☐ ☐ ☐

46 What are isotopes? ☐ ☐ ☐

The periodic table

47 What does the periodic table contain? ☐ ☐ ☐

48 What are the columns called? ☐ ☐ ☐

49 What are the rows called? ☐ ☐ ☐

Metals

50 Where are the metals found in the periodic table? ☐ ☐ ☐

51 Give three properties of metals. ☐ ☐ ☐

52 What is an alloy? ☐ ☐ ☐

Group I

53 What are the names and symbols of the first three elements in group I? ☐ ☐ ☐

54 What are the two products when sodium reacts with water? ☐ ☐ ☐

55 Name the product when potassium reacts with chlorine. ☐ ☐ ☐

Transition metals

56 Where are the transition metals in the periodic table? ☐ ☐ ☐

57 What does 'malleable' mean? ☐ ☐ ☐

58 What does 'ductile' mean? ☐ ☐ ☐

59 Give one property of transition metal compounds. ☐ ☐ ☐

Metals and their reactivity

60 What are the two non-metals in the reactivity series? ☐ ☐ ☐

61 What is a displacement reaction? ☐ ☐ ☐

62 What gas is produced when zinc reacts with an acid? ☐ ☐ ☐

Answers 43 number of protons/electrons. 44 number of protons and neutrons in nucleus. 45 2 in first shell, 8 in each of 2nd and 3rd shells. 46 atoms of element with same number of protons but different numbers of neutrons. 47 all known elements 48 groups 49 periods 50 left-hand side and in central block. 51 Conductors of heat and electricity, high melting points. 52 mixture of metals. 53 lithium Li, sodium Na and potassium K 54 sodium hydroxide and hydrogen 55 potassium chloride 56 between groups II and III 57 can be made into different shapes 58 can be made into wires 59 high density; good conductors of heat and electricity; high melting points. 60 carbon (below aluminium; above zinc) hydrogen (above copper; below lead) 61 more reactive metal displaces less reactive metal from a compound. 62 hydrogen

The non-metals

63 What is group VII called?	☐ ☐ ☐
64 Which is the most reactive halogen?	☐ ☐ ☐
65 Why are group VIII elelments called inert gases?	☐ ☐ ☐
66 What is helium used for?	☐ ☐ ☐

Bonding

67 What happens to electrons in an ionic bond?	☐ ☐ ☐
68 Give three properties of ionic compounds.	☐ ☐ ☐
69 What happens to electrons in covalent bonding?	☐ ☐ ☐
70 Give two properties of covalent compounds.	☐ ☐ ☐

Chemical reactions

71 What happens in a decomposition reaction?	☐ ☐ ☐
72 What gas does a substance react with in a combustion reaction?	☐ ☐ ☐

Writing equations

73 Write a word equation for methane reacting with oxygen.	☐ ☐ ☐
74 Write a symbol equation for magnesium reacting with oxygen.	☐ ☐ ☐

Chemical calculations

75 What is the percentage of carbon in methane CH_4 (R.A.M. C=12 H=1)?	☐ ☐ ☐
76 What is the mass of two moles of sodium hydroxide NaOH (R.A.M. Na=23 O=16 H=1)?	☐ ☐ ☐

Answers 63 The halogens. 64 fluorine 65 Because they do not react. 66 Balloons and airships. 67 They are transferred. 68 High melting points, conduct electricity when molten or in solutions, soluble in water. 69 Electrons are shared. 70 Low melting points, do not conduct electricity. 71 Compounds are broken down 72 Oxygen 73 Methane + oxygen → carbon dioxide + water 74 $2Mg + O_2$ → $2MgO$ 75 $12/16 \times 100 = 75\%$ (R.A.M. C=12 H=1) 76 Mass of 1 mole of NaOH = 23 + 16 + 1 = 40 Mass of 2 moles = $2 \times 40 = 80$ g (R.A.M. Na=23 H=1 O=16)

Acids and their reactions

77 What is the pH of an acid?	☐ ☐ ☐
78 What ions do all acids contain?	☐ ☐ ☐
79 What are two products are formed when magnesium reacts with hydrochloric acid?	☐ ☐ ☐
80 What are the three products formed when calcium carbonate reacts with sulphuric acid?	☐ ☐ ☐

Bases and neutralisation

81 What is the difference between a base and an alkali?	☐ ☐ ☐
82 What two types of substances react together in a neutralisation reaction?	☐ ☐ ☐

Rate of reaction I

83 What four factors effect rates of reaction?	☐ ☐ ☐
84 What does the collision theory state?	☐ ☐ ☐

Rate of reaction II

85 What is the difference between exothermic and endothermic reaction?	☐ ☐ ☐
86 What is meant by a reversible reaction?	☐ ☐ ☐

Enzymes

87 What are enzymes?	☐ ☐ ☐
88 What is produced in fermentation?	☐ ☐ ☐
89 Give three uses of enzymes in industry.	☐ ☐ ☐

Useful products from oil

90 How are the parts of oil separated?	☐ ☐ ☐
91 What elements do hydrocarbons contain?	☐ ☐ ☐
92 What is cracking?	☐ ☐ ☐

Answers 77 Between pH 1 and 6. 78 Positive hydrogen ions.
79 Magnesium chloride and hydrogen. 80 Calcium sulphate,
carbon dioxide and water. 81 An alkali is a base which
dissolves in water. 82 An acid and a base or alkali.
83 Temperature, concentration, surface area and a catalyst.
84 That particles must collide with sufficient energy for the
reaction to take place. 85 Exothermic reactions give out heat
and endothermic reaction take in heat. 86 A reaction that
can go both ways. 87 biological catalysts 88 alcohol and
carbon dioxide 89 stone-washing denim, in baby food, in
slimming food 90 by fractional distillation. 91 Hydrogen and
carbon. 92 Breaking down large hydrocarbons into smaller
ones.

Ammonia and fertilizers

93 What is the formula for ammonia?	☐	☐	☐
94 What are the raw materials used in the Haber process?	☐	☐	☐
95 What do all fertilisers contain?	☐	☐	☐
96 What is the process called when excess fertiliser goes into a river?	☐	☐	☐

Extraction of metals

97 By what process is aluminium extracted from its ore?	☐	☐	☐
98 What is the name of the ore of aluminium?	☐	☐	☐
99 What happens to the carbon anodes in this process?	☐	☐	☐
100 What is haematite?	☐	☐	☐
101 What is the reducing agent in the blast furnace?	☐	☐	☐
102 What is slag?	☐	☐	☐
103 How is copper purified?	☐	☐	☐

Rocks and the rock cycle

104 Name the 3 types of rocks.	☐	☐	☐
105 What are the 3 types of weathering?	☐	☐	☐
106 How are sedimentary rocks changed to metamorphic rocks?	☐	☐	☐

Changes in the atmosphere

107 What were the main gases produced when the Earth was formed?	☐	☐	☐
108 What gas did the plants start to produce?	☐	☐	☐
109 What percentage of nitrogen is there in the earth's atmosphere today?	☐	☐	☐
110 What gas causes global warming?	☐	☐	☐

Answers 93 NH_3 94 Air, water (steam) and methane.
95 Nitrogen 96 Eutrophication 97 Electrolysis
98 bauxite 99 They react with the oxygen produced to form carbon dioxide and this means the electrodes burn away.
100 Iron ore. 101 Carbon monoxide. 102 The impurities in the blast furnace which reacts with the limestone.
103 By electrolysis. 104 Igneous, sedimentary and metamorphic. 105 Biological, physical and chemical.
106 They are changed by heat and pressure. 107 Carbon dioxide, ammonia and methane 108 Oxygen 109 78% 110 Carbon dioxide

Electricity and magnetism

111 What is the relationship between current I, charge Q and time t?	□	□	□
112 What is the relationship between voltage V, current I and resistance R?	□	□	□
113 What is the relationship between power P, voltage V and current I?	□	□	□
114 What are the units used to measure current, voltage, resistance, and power?	□	□	□
115 What is the direction of the electrical force when objects with the same charge are brought close together?	□	□	□
116 How can you change the direction and size of the force on a current-carrying coil of wire in a magnetic field?	□	□	□
117 What is electromagnetic induction?	□	□	□

Force and motion

	□	□	□
118 What is the relationship between speed, distance travelled and time taken?	□	□	□
119 What is the relationship between speed, time and acceleration?	□	□	□
120 What can you say about the motion of an object when the forces on the object are balanced?	□	□	□
121 What can you say about the forces on an object that is accelerating?	□	□	□
122 How does the acceleration of an object depend on its mass?	□	□	□
123 What are the factors that affect the stopping distance of a car?	□	□	□
124 What is the difference between mass of an object and its weight, and what are their units?	□	□	□

Answers 111 $Q = I\,t$ 112 charge = current $\times$ time ;
113 $V = I\,R$ 114 power = voltage $\times$ current 115 current –
amperes (A), voltage – volts (V), resistance – ohms (Ω) and
power watts (W). 116 like charges repel 117 Change the
direction of the force by changing the direction of the current
OR the direction of the magnetic field. Increase the size of the
force by increasing the current, OR by increasing strength of
the magnetic field OR by increasing the number of turns of
wire on the coil.

118 Electromagnetic induction is the production of a voltage
across a wire that is cutting across a magnetic field.
119 speed $= \frac{\text{distance travelled}}{\text{time}}$ 120 acceleration $= \frac{\text{change in speed}}{\text{time taken}}$
121 If the forces are balanced the object is either stationary
or moving at a steady speed 122 The forces are unbalanced.
123 The bigger the mass of an object the smaller the
acceleration when the force remains constant. 124 Mass is
measured in kilograms, weight is measured in newtons and
includes gravity as a component.

Waves

125 What is the difference between transverse and longitudinal waves – give an example of each.	☐ ☐ ☐
126 What is the relationship between the frequency f, the speed v and the wavelength λ of a wave?	☐ ☐ ☐
127 What is refraction of waves?	☐ ☐ ☐
128 Put these components of the electromagnetic spectrum into order of wavelength, starting with the shortest: gamma rays, infrared, microwaves, radio waves , ultraviolet, visible light, x-rays	☐ ☐ ☐
129 Write down a use for each of the components of the spectrum listed in 128.	☐ ☐ ☐
130 What is ultrasound and how is it used?	☐ ☐ ☐
131 What are the differences between P earthquake waves and S earthquake waves?	☐ ☐ ☐

Earth and beyond

132 List the planets in the solar system in order, starting from the planet nearest to the Sun.	☐ ☐ ☐
133 What is the force that keeps the planets in orbit around the Sun?	☐ ☐ ☐
134 List three different uses of satellites.	☐ ☐ ☐
135 How are stars formed?	☐ ☐ ☐

Answers 125 In transverse waves the oscillation is at right angles to the direction the wave travels – e.g. light, water waves. In longitudinal eaves the oscillation is in the same direction that the wave travels – e.g. sound waves. 126 v = fλ 127 Refraction occurs when a wave passes from one medium to another, the speed changes and the wave changes direction e.g. when light passes from air to glass. 128 gamma rays, x-rays, ultraviolet, visible light, infrared, microwaves, radio waves 129 gamma rays used in treatment of cancer; x-rays used for medical imaging, ultraviolet used for sterilisation, visible light used for seeing, infrared used in communications, microwaves used for cooking and communications, radio waves used for communications 130 Ultrasound is sound of frequencies too high to be heard. Used for imaging and also to measure distances. 131 P waves are longitudinal waves; S waves are transverse waves. P waves travel more quickly than S waves. 132 Mercury, Venus, Earth, Mars, Jupiter, Saturn, Uranus, Neptune, Pluto 133 gravity 134 Communications, observing the surface of the Earth and observing the Universe 135 A cloud of dust and gases is drawn together by gravity, when the forces between the materials are large enough and the material gets hot enough nuclear fusion takes place and light is given out.

Work, power and energy

136 What is the relationship between force, work and distance?	☐ ☐ ☐
137 What is the relationship between power, energy and time?	☐ ☐ ☐
138 What are the units used to measure energy, force, work, distance, power and time?	☐ ☐ ☐
139 What equations are used to calculate kinetic energy and change in gravitational potential energy?	☐ ☐ ☐
140 What are the ways in which energy can be transferred from a hot object to its surroundings?	☐ ☐ ☐

Radioactivity

141 What are three components of an atom?	☐ ☐ ☐
142 What are the properties of alpha particles?	☐ ☐ ☐
143 What are the properties of beta particles?	☐ ☐ ☐
144 What are the properties of gamma rays?	☐ ☐ ☐
145 What is meant by the half-life of a radioactive element?	☐ ☐ ☐

136 work = force $\times$ distance moved in the direction of the force 137 power = $\frac{\text{work done or energy transferred}}{\text{time taken}}$ 138 energy – joule (J); force – newton (N); work – joule (J); distance – metre (m); power – watt (W); time – second (s). 139 change in gravitational potential energy = m g h; kinetic energy = $\frac{1}{2}$ mv^2 140 conduction, convection, radiation and evaporation 141 electrons, protons and neutrons 142 Alpha particles are made of 2 protons and 2 neutrons; they are positively charge and stopped by paper. 143 Beta particles are electrons and are negatively charged, they are stopped by a few millimetres of aluminium 144 Gamma rays are electromagnetic waves; they are stopped by several centimetres of lead or a greater thickness of concrete. 145 The half-life of a radioactive element is the time it takes for half the radioactive nuclei in a sample of the material to decay.

1   Complete the labels for the diagram and the corresponding phrases next to it.

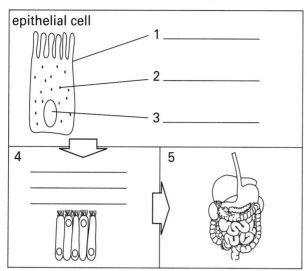

1 The cell membrane controls

the _____.

2 The cell contents, called

_____, are where chemical

reactions occur.

3 The _____ contains

genetic material.

4 Many epithelial cell make up a

_____ called the

epithelium, which lines the gut.

5 The stomach is an _____

within the _____ system.

2   Complete the flow chart of digestion:

main food types:	
egg = _____	butter = _____
bread = _____	

In the mouth	In the stomach	In the small intestines
_____	_____	_____ the remaining
is broken down	is broken down	food is digested, and
by _____	by _____	_____ into the
into _____.	Into _____.	bloodstream.

3   Complete the notes on the graph which shows the rate at which an enzyme works.

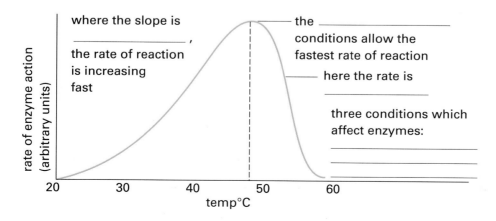

where the slope is

_____ ,

the rate of reaction
is increasing
fast

the _____
conditions allow the
fastest rate of reaction

here the rate is

_____

three conditions which
affect enzymes:

_____
_____
_____

4 Fill in the blanks on the heart diagram (below) about its structure, and how it beats.

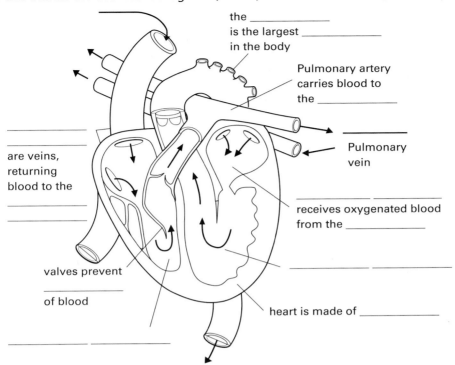

the _____
is the largest _____
in the body

Pulmonary artery
carries blood to
the _____

_____
are veins,
returning
blood to the
_____
_____

Pulmonary
vein

_____ _____
receives oxygenated blood
from the _____

_____ _____

valves prevent
_____
of blood

heart is made of _____

_____ _____

5 Complete the table on microbes.

name	example	disease it causes	medicines used to fight infection
virus			
bacterium			
fungus			

6 Complete the sentences about immune response:

An _____ is a particle or cell, which is foreign to the body. These are detected

by _____ _____ cells, which produce _____ to match. The _____

lock onto the antigens, damaging them or allowing white blood cells to _____

them. Some white blood cells make _____, which can neutralise toxins.

7 Complete the table of breathing events, choosing from the list in the last column.

	breathing in	breathing out	answers
ribs move			up/down
diaphragm			up /down
volume of chest			increases/decreases
air pressure inside chest			increases/decreases
air flows			out/in

8 Complete the equation for respiration.

glucose + _____ → _____ + _____ [+ _____ transferred]

9 Complete the phrases in the flowchart about the main parts of the nervous system.

detecting   processing   responding

A _____ neurone has a _____ to detect a stimulus. Bundles of these neurones form a _____, which carries impulses to the _____ nervous system.

The _____ and _____ _____ are the main parts of the central nervous system. A reflex action may not involve the _____.

_____ neurones lead from the central nervous system to an _____ e.g. a _____ which produces a hormone or _____, which causes movement.

10 Complete the table about the components of the eye and their functions.

part of the eye	function
cornea	
iris	
pupil	
eyelash	
retina	
blood supply	
tear gland	

11 Complete the following sentences:

Insulin is the hormone which _____

Glycogen is a _____

Glucagon causes _____

Blood sugar level changes because _____

12 Write a definition for each of the following terms:

ovulation	menstruation	implantation	fertilisation
_____	_____	_____	_____
_____	_____	_____	_____
_____	_____	_____	_____
_____	_____	_____	_____

13 Name one male and two female sex hormones.

male: _____   female: _____   _____

14 Complete the table about the male and female reproductive systems.

part of the body	function
	produces sperm
	fertilised eggs develop
scrotum	
	fuses with the egg cell during fertilisation
urethra	

15 Complete the flowchart describing what happens when Tim works out at the gym:

Tim works out with weights, increasing _____ in the muscles. This transfers more energy and body temperature _____ .

➤

Tim needs to mop his forehead because _____ production increases. His face gets _____ because of increased _____ _____ near the skin surface.

➤

Small _____ near the skin surface widen, which is called _____ . Heat _____ from the body, causing cooling.

16 Name the organs responsible for adjusting water content of the body.

_____

17 Fill in the blanks to complete the labels.

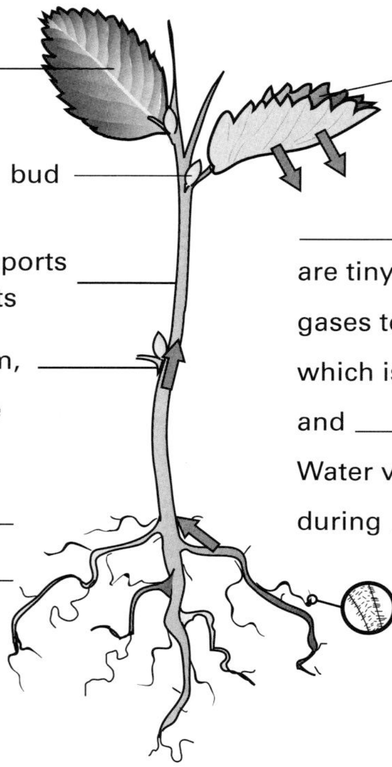

_____ is the green material in leaves which absorbs light energy.

the surface of the leaf has a waxy _____, which helps to _____ water loss.

bud

stem supports the shoots

_____ are tiny pores in the leaf surface which allow gases to move in and out eg. _____, which is a raw material for photosynthesis, and _____, which is a by-product. Water vapour also passes out of the leaf during _____ .

Water moves up the stem, supplying all parts of the shoot system. This is a _____ for photosynthesis

Tiny _____ _____ increase _____ for absorption of water.

18 Complete the following sentences about plant hormones, by filling the blanks.

_____ are plant hormones, which affect the growth of cells, by affecting the _____ of cell division and the amount of cell _____. In _____, the rate of cell growth increases where there is a higher concentration of hormone. This causes uneven growth on either side and a change of direction. Hence a shoot may grow towards light. This type of growth movement is called a _____.

19 Complete the table about types of reproduction by writing true (T) or false (F) in a box.

	asexual	sexual
two parent cells are involved	☐	☐
mitosis is the type of cell division which occurs	☐	☐
cloning is an example of this type of reproduction	☐	☐
there is genetic variation	☐	☐
sex cells are produced	☐	☐

20 The people named below developed theories of evolution. Match statements which are true for each person's theory by drawing a line from the statement to the name.

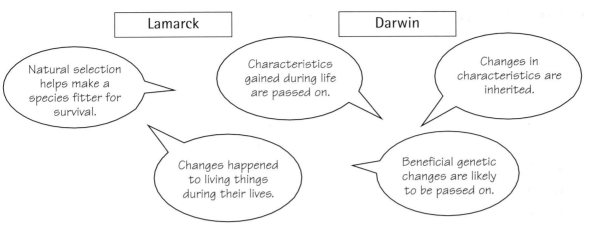

Lamarck   Darwin

Natural selection helps make a species fitter for survival.

Characteristics gained during life are passed on.

Changes in characteristics are inherited.

Changes happened to living things during their lives.

Beneficial genetic changes are likely to be passed on.

21 Complete the Punnet square, showing the inheritance of red flower colour (R) and white (r) flower colour.

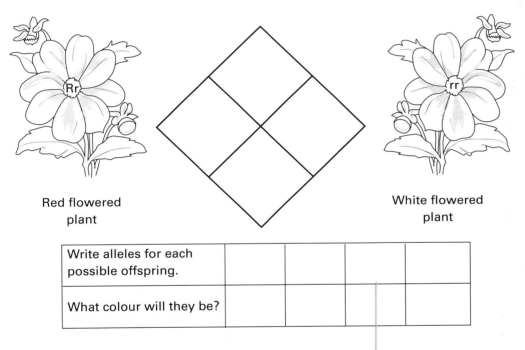

Red flowered plant

White flowered plant

Write alleles for each possible offspring.				
What colour will they be?				

22 Name five types of pollution, which may be caused by human activities.

_____   _____   _____

_____   _____

23 Write a food chain which might occur in

a woodland   _____ → _____ → _____ → _____

the desert   _____ → _____ → _____ → _____

the sea   _____ → _____ → _____ → _____

24 The pyramid of biomass represents food webs from African grasslands. Complete the labels by filling the blanks.

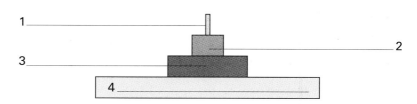

25 Write a definition for the terms:

predator _____

and *prey* _____

26 Are these statements true or false? Write true or false next to each one.

Carbon dioxide enters the atmosphere as a result of: a) respiration _____

b) transpiration _____

Oxygen is a by-product of: a) fermentation _____

b) photosynthesis _____

Burning fossil fuels produces: a) carbon dioxide _____

b) nitrogen oxide _____

Chemistry

1 Label the diagram of the atom:

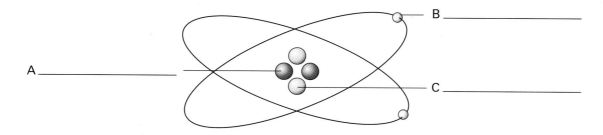

A _____

B _____

C _____

2 Complete the following table:

element	atomic number	mass number	number of protons	number of electrons	number of neutrons
carbon	6	12			
fluorine			9		10
hydrogen	1	1		1	0
iron				26	30

3 In the periodic table below:
- highlight in red the alkali metals
- highlight in blue the halogens
- highlight in yellow the noble gases
- highlight in orange the transition metals.

							H Hydrogen 1										He Helium 2
Li Lithium 3	Be Beryllium 4											B Boron 5	C Carbon 6	N Nitrogen 7	O Oxygen 8	F Fluorine 9	Ne Neon 10
Na Sodium 11	Mg Magnesium 12											Al Aluminium 13	Si Silicon 14	P Phosphorus 15	S Sulphur 16	Cl Chlorine 17	Ar Argon 18
K Potassium 19	Ca Calcium 20	Sc Scandium 21	Ti Titanium 22	V Vanadium 23	Cr Chromium 24	Mn Manganese 25	Fe Iron 26	Co Cobalt 27	Ni Nickel 28	Cu Copper 29	Zn Zinc 30	Ga Gallium 31	Ge Germanium 32	As Arsenic 33	Se Selenium 34	Br Bromine 35	Kr Krypton 36
Rb Rubidium 37	Sr Strontium 38	Y Yttrium 39	Zr Zirconium 40	Nb Niobium 41	Mo Molybdenum 42	Tc Technetium 43	Ru Ruthenium 44	Rh Rhodium 45	Pd Palladium 46	Ag Silver 47	Cd Cadmium 48	In Indium 49	Sn Tin 50	Sb Antimony 51	Te Tellurium 52	I Iodine 53	Xe Xenon 54
Cs Caesium 55	Ba Barium 56	La Lanthanum 57	Hf Hafnium 72	Ta Tantalum 73	W Tungsten 74	Re Rhenium 75	Os Osmium 76	Ir Iridium 77	Pt Platinum 78	Au Gold 79	Hg Mercury 80	Tl Thallium 81	Pb Lead 82	Bi Bismuth 83	Po Polonium 84	At Astatine 85	Rn Radon 86
Fr Francium 87	Ra Radium 88	Ac Actinium 89															

4 Draw in the dots and crosses to show the bonding in the following substances:

a) H O H   b) Cl   Cl   c) H   N   H

H

5 Match the word equations with the type of reaction:

combustion	i) sodium + sulphuric → sodium sulphate + water hydroxide acid

neutralisation	ii) magnesium + iron → magnesium + iron sulphate sulphate

displacement	iii) methane + oxygen → carbon dioxide + water

decomposition	iv) calcium carbonate → calcium oxide + carbon dioxide

6 Complete the following word equation:

Hydrochloric acid + zinc → _____

Sulfuric acid + sodium carbonate → _____

Nitric acid + copper oxide → _____

Hydrochloric acid + potassium hydroxide → _____

7 Add the following labels to the pH scale (below):

neutral, weak acid, weak alkali, strong acid, strong alkali

1   2   3   4   5   6   7   8   9   10   11   12   13   14

a)_____   b)_____   c)_____   e)_____

d)_____

Red   Orange   Yellow Green   Blue   Purple

8 Complete the following table about enzymes

use of enzyme	brief explanation
fermentation	
yoghurt making	
stone-washing denim	
in slimming foods	

9 Complete the following requirements for the Haber process:

raw material for nitrogen: _____

raw materials for hydrogen: _____ and _____

catalyst:_____

temperature:_____

pressure:_____

10 Complete the diagram of the purification of copper

11 Match the type of rock to the correct examples

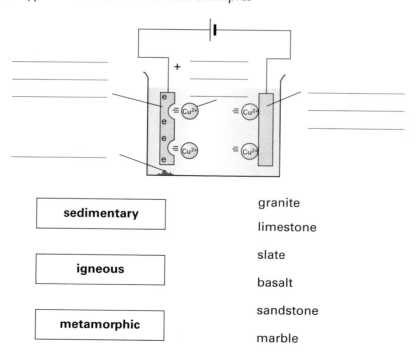

sedimentary	granite
	limestone
	slate
igneous	basalt
	sandstone
metamorphic	marble

Physics

1 Use the variables in the centre to make equations:

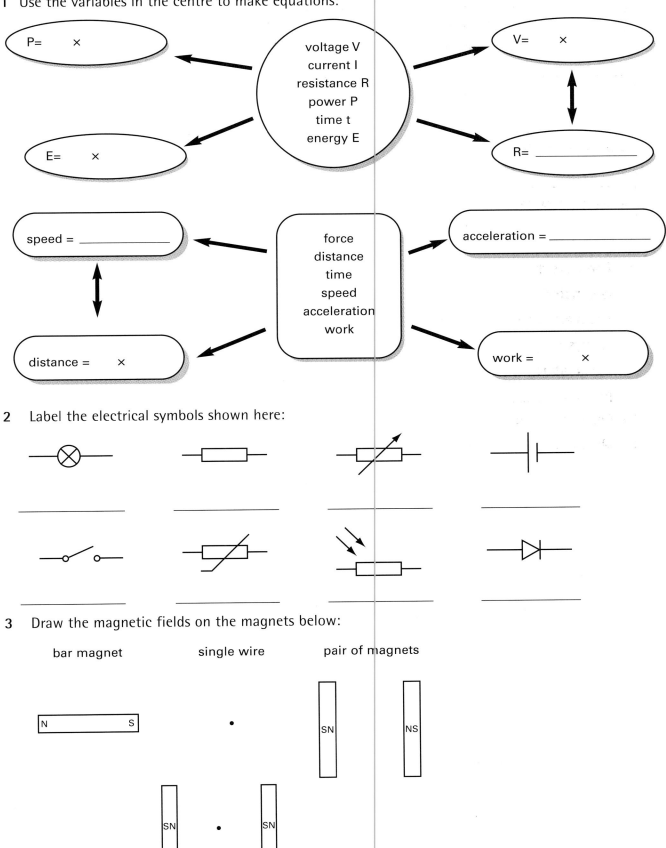

voltage V
current I
resistance R
power P
time t
energy E

P= ×

V= ×

E= ×

R= _____

force
distance
time
speed
acceleration
work

speed = _____

acceleration = _____

distance = ×

work = ×

2 Label the electrical symbols shown here:

_____ _____ _____ _____

_____ _____ _____ _____

3 Draw the magnetic fields on the magnets below:

bar magnet   single wire   pair of magnets

N   S

SN   NS

SN   SN

single wire in magnetic field

4 Complete the sentences on the following transformers:

The primary coil of a transformer is connected to an _____ current supply.

As the current in the primary coil varies it sets up a _____ _____ field in the

iron core, which _____ a changing voltage in the secondary coil.

5 Fill in the gaps in the tables below on waves:

electromagnetic	gamma rays		ultraviolet		infrared		radio waves
uses		medical imaging		seeing		cooking and communications	

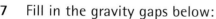

wave	oscillation direction	examples
longitudinal		
transverse		

6 Name the planets in order from the Sun:

Mercury				asteroid belt					Pluto

7 Fill in the gravity gaps below:

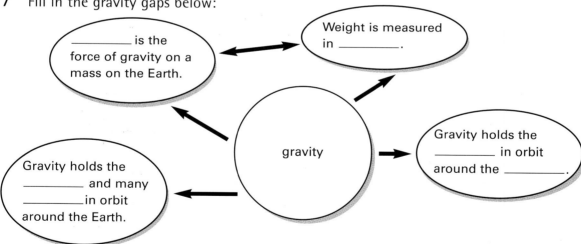

8 Complete the life story of a star:

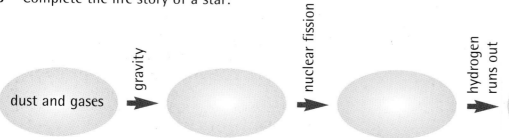

9 Complete the table on energy transfer:

method of transfer	what happens in the material	example
	energy of vibrating particles is transferred to neighbouring particles	metals are good conductors and are used to make saucepans
convection		
radiation		energy reaches from the sun this way
		sweating removes energy and prevents the body from overheating

10 Fill in the gaps about radioactivity:

radiation	what is it?	charge?	how far can it travel in air?	what stops it?
alpha α				
beta β				
gamma γ				

Biology answers

1 1 movement in/out of cell   2 cytoplasm
3 nucleus 4 tissue   5 organ, digestive

2 egg: protein, butter: fat/lipid, bread: carbohydrate
starch, carbohydrase/amylase, sugar/glucose;
protein, protease, amino acids; all, absorbed

3

where the slope is steepest, the rate of reaction is increasing fast

the optimum conditions allow the fastest rate of reaction

here the rate is decreasing

three conditions which affect enzymes: PH, temperature, substrate concentration.

4 starting top right: aorta; artery; lungs;
culmonary vein; right atrium/auricle; lungs;
left ventricle; muscle; right ventricle; backflow;
venacava; left atrium

5

name	example	disease	medicines
virus	HIV	AIDS	anti-virals
bacterium	streptococcus	sore throat	antibiotic
fungus	various	athletes foot	antifungal

6 antigen, white blood, antibodies, antibodies, engulf, antitoxins

7

breathing in	breathing out
up	down
down	up
increases	decreases
decreases	increases
out	in

8 glucose + oxygen → carbon dioxide + water
[+ energy transferred]

9 A neurone has a sensory nueron to detect a stimulus.
Bundles of these neurones form a nerve, which carries impulses to the central nervous system.

The brain and spinal cord are the main parts of the central nervous system.
A reflex action may not involve the brain.
Motor neurones lead from the central nervous system to an effector e.g. a gland, which produces a hormone or muscle, which causes movement.

10 cornea: helps focus light/form image
iris: controls amount of light entering eye
pupil: allows light to enter eye
eyelash: traps dirt/stops dust getting in eye
retina: detects light/sends impulses to brain
blood supply: supplies eye with oxygen
tear gland: washes eye with tears/prevents infection/antibacterial

11 Insulin: converts glucose to glycogen; Glycogen: carbohydrate store; Glucagon: conversion of glycogen to glucose; intake/what we eat varies

12 ovulation: release of eggs from ovary; menstruation: monthly loss of blood; implantation: fertilised embryo becomes attached/embeds in lining of womb; fertilisation: nucleus of two sex cells/sperm and egg fuse

13 male: testosterone
female: FSH, oestrogen and progesterone.

14 testis: produces sperm
uterus: fertilised eggs develop
scrotum: supports testes
nucleus of sperm: fuses with the egg during fertilisation
urethra: passes urine from bladder out of body

15 Tim works out with weights, increasing respiration in muscles. This transfers more energy and body temperature increases. Tim needs to mop his forehead because sweat production increases. His face gets flushed/red because of increased blood circulation near the skin surface. Small capillaries near the skin surface widen, which is called vasodilation. Heat radiates from the body, causing cooling.

16 kidneys

17 clockwise: cuticle; reduce; stomata; carbon dioxide; oxygen; transpiration; root; hairs; surface; material; raw; chlorophyll.

18 auxins, rate, elongation/growth, stems/shoot tips, tropism

19 sexual; asexual; asexual; sexual; sexual

20 Natural selection helps make a species fitter for survival. (Darwin); Characteristics gained during life are passed on. (Lamarck); Changes in characteristics are inherited. (Darwin); Beneficial genetic changes are likely to be passed on. (Darwin); Changes happen to living things during their lives. (Lamarck)

21
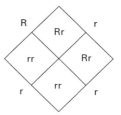

Write alleles for each possible offspring.	Rr	Rr	rr	rr
What colour will they be?	Red	Red	White	White

22 increase carbon dioxide in air; acid rain; chemicals/sewerage in waterways; PCBs in air; garbage/landfill; nuclear waste

23 **woodland:** tree → squirrel → fox → decomposers;
desert: cactus → insect → lizard → vulture;
sea: plankton → small fish → large fish → humans

24
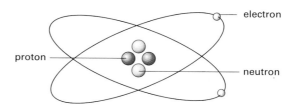

25 predator: animal that hunts/catches and eats other animals; prey: animal that is hunted/eaten by other animals

26 true, false, false, true, true, true

Chemistry answers

1
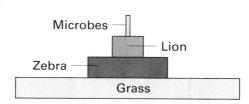

2

atomic number	mass number	number of protons	number of electrons	number of neutrons
6	12	6	6	6
9	19	9	9	10
1	1	1	1	0
26	56	26	26	30

3

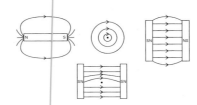

4 a)
H $\overset{\times\ \times}{\underset{\times\ \times}{\overset{\times}{\underset{\bullet}{O}}}}$ H

b)
$\overset{\bullet\bullet}{\underset{\bullet\bullet}{Cl}}\ \overset{\times\ \times}{\underset{\times\ \times}{Cl}}$

c)
H $\overset{\times\ \times}{\underset{\times}{\overset{\bullet\bullet}{N}}}$ H
$\quad$ H

5 Combustion: iii
Neutralisation: i
Displacement: ii
Decomposition: iv

6 = zinc chloride + hydrogen
= sodium sulphate + carbon dioxide + water
= copper nitrate + water
= potassium chloride + water

7 a) strong acid b) weak acid c) weak alkali
d) neutral e) strong alkali

8 **fermentation:** used for making beer, bread
and wine. Yeast is used with sugar to produce
alcohol and carbon dioxide.
yoghurt-making: enzymes convert sugar in
milk to lactic acid to make proteins thicken.
stone-washing denim: add enzyme to denim,
which will stonewash denim rather than adding
stones.
in slimming foods: enzymes changes glucose to
fructose, which is sweeter.

9 **nitrogen:** air; **hydrogen:** water (steam) and
methane; **catalyst:** iron; **temperature:** 450
–500°C; **pressure:** 200 Atm

10

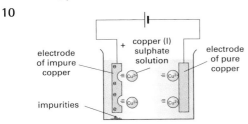

electrode of impure copper

+ copper (I) sulphate solution

electrode of pure copper

impurities

11 sedimentary: sandstone and limestone, igneous:
basalt and granite, metmorphic: marble and slate.

Physics answers

1 $P = V \times I$; $V = I \times R$; $E = P \times t$; $R = V / I$
speed = distance ÷ time; acceleration = change
in speed ÷ time taken; work = force × distance;
distance = speed × time

2 A: lamp, B: resistor, C: variable resistor, D: cell,
E: switch, F: thermistor, G: LDR, H: diode.

3

4 The primary coil is connected to an <u>alternating</u>
current supply. As the current in the primary
coil varies it sets up a <u>changing magnetic</u> field
in the iron core, which <u>induces</u> a changing
voltage in the secondary coil.

5 **gamma rays:** treatment of cancer; **x-rays:**
medical imaging; **ultraviolet:** sterilising medical
equipment; **visible light:** seeing; **infrared:**
heating; **microwave:** cooking and
communications; **radio waves:** communications.
longitudinal wave: oscillates in same direction as
energy oscillates (e.g. sound waves, earthquake p
waves); transverse waves: travels at right angles
to direction energy travels (e.g. light, water,
earthquake S waves)

6 Mercury, Venus, Earth, Mars, asteroid belt,
Jupiter, Saturn, Uranus, Neptune, Pluto.

7 **Weight** is the force of gravity on a mass on the
Earth; Weight is measured in **newtons**; gravity
holds the **planets** in orbit around the Sun;
Gravity holds the **Moon** and many **satellites** in
orbit around the Earth.

8 very hot gases, adult star, red giant

9 **conduction:** energy of vibrating particles is
transferred to neighbouring particles
: metals are good conductors and are used to
make saucepans
convection: warmer, less dense liquid or gas
rises above denser, older fluid
: energy from the element at the bottom of a
kettle is transferred to all the water.
radiation: electromagnetic waves (infrared)
: energy reaches us from the Sun this way.
evaporation: particles evaporating from a liquid
carry away energy, leaving the liquid cooler
: sweating removes energy and prevents the
body from overheating.

10

alpha α	2 protons 2 neutrons	positive	few centimetres	paper
beta β	electron	negative	few metres	aluminium
gamma γ	electro-magnetic wave	none	a long way	thick lead, very thick concrete

Answers

Life processes

A   awareness of danger, selecting better environments, finding food
A   many different cell types adapted for different functions
1   permanent size increase, due to increase in size/number of cells.
2   feeding/nutrition
3   to transfer energy form food to cells/body; to get energy
4   plants make own food by photosynthesis, animals must eat ready-made food
5   light sensitive cell, nerve tissue, the eye

Cells and cell structure

A   nucleus acts as a set of instructions for cell processes, including inheritance of characteristics; chloroplast is where photosynthesis happens in plant cells; surface membrane controls what moves in and out of cells
A   job is to carry out photosynthesis, it contains many chloroplasts in order to do this
A   reach a maximum size; more cells are produced
1   nucleus
2   cellulose cell wall and vacuole containing watery sap
3   both involved in absorption/exchange of materials
4   chloroplast

Body systems

A   chloroplasts
A   to transport a wide variety of materials within the organism
1   leaves get more light if they grow above other plants
2   up through transporting tissue, from root to leaves
3   flower
4   blood system
5   blood system
6   nervous system – sends impulses; muscles move in response; blood transports food and oxygen to muscles

Human nutrition

A   sugar/glucose
A   don't eat enough fresh fruit and vegetables in diet
A   generally larger body mass/more muscle tissue
1   fat/lipid; carbohydrate (sugar/starch)
2   need energy, contains essential components
3   contain more vitamins than highly cooked foods; contain fibre
4   protein
5   beans e.g. soya, tofu; pulses e.g. lentils; nuts e.g. peanuts
6   less active so need less energy
7   eating too much, eating high calories foods e.g. fat and sugar, lack of exercise

Human digestion

A   liver
A   peristalsis
A   pH in stomach is too low/too acid
1   stomach
2   salivary gland; pancreas; lining of the small intestine
3   long length; inner surface has finger-shaped villi; cell surfaces have microvilli
4   absorbed into bloodstream/capillaries of gut wall
5   shape of protein changes/active site destroyed

The heart

A   left atrium, right atrium, left ventricle, right ventricle, aorta, vena cava
A   no need to think about it, happens when we sleep, can't forget to breathe, better regulation
1   stronger muscle, pumps a larger circulation
2   contracts in a wave/atria contract then ventricles/squeeze blood out into main arteries
3   valves between atria and ventricles close when ventricles are full and begin to contract; valves at base of large arteries close when ventricles stop contracting

Circulation and the blood system

A   one circulation goes to the lungs and returns to the heart; the other pumps around the rest of the body
A   red blood cells, plasma
1   aorta receives blood at higher pressure as left ventricle contracts, while vena cava collects blood, which is some distance from the heart; aorta contains oxygenated blood from the left side of the heart, but blood in vena cava has passed through tissues and oxygen has diffused into cells
2   diffusion gradient/difference in concentration of oxygen

Health and disease

A   unable to carry out normal activities due to symptoms
A   virus, bacterium, fungus
A   infection, injury, lifestyle habit, mental illness, inherited
1   an infection
2   immunisation
3   produce antibodies; engulf/destroy germs/microbes/bacteria; neutralise toxins

Immunity

A   injury can result in bleeding to death
A   white blood cells
1   diet, exercise, not abusing drugs (including not smoking)
2   incidence of lung cancer is greater for smokers than non-smokers

The breathing system

A   trachea – bronchi – bronchioles
A   active process/muscles contract to lift ribs/flatten diaphragm
A   tars; nicotine; carbon monoxide
1   many of them, large surface area, thin-walled for fast diffusion, rich blood supply
2   by changing volume of chest cavity, which changes air pressure inside lungs compared to outside
3   nicotine causes blood vessels to contract

Transferring energy by respiration

A   to transfer energy to cells
A   muscles need energy, use muscles more when exercising
A   for chemical processes e.g. building new molecules; allow muscles to move; heat keeps body warm
1   carbon dioxide
2   aerobic uses oxygen, anaerobic doesn't; aerobic transfers more energy than anaerobic; glucose completely broken down to CO_2 and water during aerobic, only partly broken down in anaerobic.

3 when there is a build up of lactate/demand for oxygen is greater than supply

4 210 kJ

The basics of sensitivity

A receptor detects changes/stimuli; effector responds/acts

A to avoid potential damage/learn from experience

1 one end is developed as a receptor/can detect stimuli/changes

2 to insulate the axon

3 interprets information and decides on action

Responding with sensitivity

A alcohol alters judgement; is a sedative: slows responses and causes drowsiness

A time taken to respond increases after drinking alcohol

1 reflex action: moving bare foot of sharp stone, crying when peeling onions, catching onto a rail to save yourself from falling; others are voluntary responses

2 Darren; his liver did not break down the alcohol as quickly

The eye and sight

A retina

A photoreceptors in retina, sensory neurones in optic nerve to brain, motor neurones to iris

1 eyelid closes over eye surface; blinking reflex; eye sunk deep into bony socket; eyelashes catch dirt

2 antibacterial tears

3 middle layer/choroid

4 helps to focus light rays

5 takes information from receptor cells to the brain

6 brain interprets the image

7 fuzzy/out of focus

Controlling body conditions carefully

A situated at base of brain/controls other glands

A medicine made by a bacterium, which contains a human gene

1 in the bloodstream

2 ovary; uterus/womb

3 beard growth, development of testes/ sperm production; body mass increases/ muscle development; pubic hair

4 pituitary gland is situated in brain

5 a) rises b) increases

6 to avoid rapid increase/decrease in blood sugar level/keep sugar intake fairly constant

Sexual maturity

A testosterone; progesterone and oestrogen

A day 14

A conception: egg fertilisation; contraception: prevents fertilisation

1 male: sperm cells; female: ova/eggs

2 lining of the uterus is lost

3 stimulate egg production

4 avoiding sexual intercourse at the time of ovulation

Body systems for reproducing

A eggs produced in ovary; fertilised in oviducts/Fallopian tubes

A sperm cells produced in the testis and stored in the epididymis

1 a: uterus, b: vagina, c: oviduct, d ovary

2 can pass through placenta to developing foetus

3 not all eggs are fertilised; fertilised eggs do not always develop

Keeping the body in balance

A water and salt balance; blood sugar levels; temperature; pH

A increase, due to greater fluid intake

A skin damage allows fluid loss, and infection

1 a) blood sugar/glucose level; b) hormone system/the liver

2 can regulate body temperature independently of environment

3 drink plenty of water to replace water lost in sweat

4 blood vessels constrict/dilate; sweat glands produce sweat

Plants and photosynthesis

A smaller raw material molecules are built into larger carbohydrate molecules

A flat/horizontal line

1 CO_2 diffuses through stomata; water absorbed by roots and is transported in xylem

2 actively growing/for cell division/building new tissues

3 supplying more CO_2 increased the rate above the 0.03% line/1.0% line is higher

4 a) summer b) winter

5 54 cm

Transport in plants

A lose water as salted water is likely to be a stronger solution than that inside cells

A organic – waste from, or remains of, living things e.g. manure, compost; inorganic – manufactured chemically in factories

A wet, still/calm air, cold (also dark)

1 stomata allow gaseous exchange

2 from an area where there is more of it/dilute solution, to where there is less of it/stronger solution

3 evaporates from cell surfaces and diffuses out of stomata

4 most of the stomata are on underside of leaf, so grease blocked them in leaf A and less water was lost

Controlling plant growth

A stems grow more rapidly until leaves reach light; help stems grow towards light by stimulating growth on shady side

A want to kill weeds e.g. dandelions, not grasses

1 shoot grow away from gravity, roots towards gravity

2 auxin is a plant hormone; chemical analysis

3 increases growth

4 inhibits side shoots

5 ripe fruit produces ethene, which causes ripening of other fruits

Variation and genetics

A each puppy inherits a slightly different combination of genes from parents' sex cells

A DNA = deoxyribose nucleic acid – chemical which makes up much of nucleus/chromosomes; chromosome is large chunk of DNA, containing many genes; gene is a smaller chunk of DNA which codes for a particular characteristic

1 during cell division when sex cells are formed; due to mutation and different combinations of individuals' genetic information

2 some characteristics are beneficial and give an organism an advantage; others are harmful or even lethal

Evolution and selection

A his ideas questioned older ideas/consider possibility of evolution

A Advantages: cheap, special equipment not essential, based on a natural process, effective over a period of time.

Disadvantages: may not produce desired effects.

1. a) 6% in unpolluted area, 53% in polluted area b) how easily they are spotted by predators/how well camouflaged they are c) decrease in dark and increase in lighter moths
2. a) high milk yield; creamy milk b) leaner meat/more meat c) rapid growth rate

Reproduction and inheritance

A 3 buds have same genes; no genetic variation; no sex cells involved

A have same parents and therefore the characteristics that come from them; different because each sex cell has a different combination of genes

A genes from engineered organisms might enter a natural population of similar organisms and alter their success

1. sexual reproduction: b, c and d; asexual reproduction: a
2. unsure of any bad effects; prefer natural foods
3. use gene to produce a medicine to compensate for the lack of a gene, or to block the effect of an undesirable gene

Inheritance

A red and white flower colour remained separate/no blending/ no pink flowers

A Bb = brown eyes; FF = freckles; XY = Male

A no chance/0% as cystic fibrosis is recessive

1. red coloured flowers appeared to hide white colour in first generation of offspring
2. R and r gives all Rr offspring (red flowered)
3. 3 red : 1 white
4. 50%

Humans and their environment

A coniferous forest, temperate forest

A increasing population

1. the environment and the living things in it
2. to preserve the gene pool
3. produces much CO_2 causes global warming

A reduction of panda population; conservation programes in protected areas; special breeding programmes.

The Earth's resources

A a) fisherman has to earn a living, and wants to catch fish while protecting stocks longer term; b) shopper is interested in choice of fish and price c) scientists want to help prevent loss of fish populations/help find ways of increasing fish stocks d) politicians have to explain management policies to the public e) farmers may farm fish; compete with fishermen for food sales

A photosynthesis

1. renewable: timber, cotton, paper, wool, soya protein
2. helps to ensure stocks for the future
3. tropical birds, e.g. macaws, turtles, terrapins – wild species.
4. adding CO_2 from burning fossil fuels, preventing deforestation, so that more carbon dioxide is removed from the atmosphere
5. increase in carbon dioxide levels causes global warming, needed by plants for photosynthesis

Surviving

A producers: lillies, bull rushes, plants; food chains e.g. plants ➜ insect ➜ frog; plants ➜ ducks; plants ➜ little fish ➜ big fish etc.; yes they do overlap

A helps to conserve water

1. population of fish which eat beetles might reduce; fish might eat more other organisms and their populations might reduce

2. plants ➜ insect ➜ frog; plants ➜ ducks; plants ➜ little fish ➜ big fish etc.
3. no natural predators for the cane toads so population increased dramatically, killing many other animals apart from rats, some of which preyed on rats
4. less water loss

Food chains and energy flow

A four

A only some biomass/energy is transferred at each level/some lost as waste

A Photosynthesis, nutrition/feeding respiration

1. a) leaves ➜ worms ➜ blackbirds ➜ foxes; or leaves ➜ beetles ➜ spiders ➜ hedgehogs ➜ foxes (or other suitable) b) producers are plants which carry out photosynthesis/make own food i.e. oak tree; primary consumer is an animal which eats a plant e.g. woodlice, beetles, worms, squirrels
2. select scale e.g. 1 cm = 10 000 kJ/m²/year; first trophic level = 88 000/10 000 or 8.8 cm, second = 1.4 cm, third = 0.16 cm and top trophic level = 0.01 cm

Atomic structure

A the electron

A 9 protons, 10 neutrons (19–9 = 10)

A 2.4

A neutrons

1. protons and neutrons
2. there are no neutrons (1–1 = 0)
3. same atomic number and same number of protons

The periodic table

A 7

A there are more metals

1. transition metals
2. sodium (Na), magnesium (Mg) and aluminium (Al)

Metals

A iron

A malleable

A iron would rust unless carbon was added to it

1. Ca, K, Cu
2. good conductors of electricity; ductile

Group I – the alkali metals

A positive ion, Na$^+$

A potassium fluoride

A potassium as it is lower in group 1

1. lithium
2. purple
3. they have 1 electron in their outer shell

Transition metals

A Fe, Zn, Cu

A a substance which will speed up a chemical reaction, without being permanently altered itself

A blue

A water and air reacting with iron

1. it is a good conductor of electricity and is ductile
2. potassium because group I metals are more reactive than transition metals
3. group I compounds: white; transition compounds: coloured

Metals and their reactivity

A potassium hydroxide and hydrogen
A no, because copper is less reactive than zinc
A hydrogen
1 Mg, Zn, Fe, Sn, Cu
2 the colour of the blue copper sulphate would fade
3 iron would displace the lead so lead and iron sulphate would be produced

Non-metals – group VII and VIII/0

A gas
A Br^-
A no chlorine is a poor conductor of electricity
A gas – they are called noble gases!
1 they have 7 electrons in their outer shell as they are in group VII
2 chlorine displaces bromine from sodium bromide, as chlorine is more reactive than bromine
3 they do not react because they have full outer shells of electrons.

Ionic bonding

A when electrons are shared or transferred between atoms
A any name, which starts with a metal and ends in a non-metal
A the ions have to be free to move
1 Na^+, Cl^-, NaCl
2 electrons
3 because the forces between oppositely-charged ions are strong forces

Covalent bonding

A covalent because it contains two non-metals
A no, as electrons are not lost or gained
A gas
A the melting point of ammonia (gas at room temperature) is vastly lower than that of diamond (solid at room temperature)
1 covalent bonds share electrons and ionic bonds involve the transfer of electrons.
2

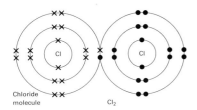

Chloride molecule Cl_2

Chemical reactions

A electrolytic decomposition
A oxygen
A a substance will: gain oxygen, lose hydrogen, lose electrons
1 oxidation (loss of an electron)
2 carbon dioxide and water
3 copper carbonate ➔ copper oxide + carbon dioxide

Writing chemical reactions

A magnesium + oxygen ➔ magnesium oxide
A NaCl
A (aq)
1 CO_2
2 $2H_2 (g) + O_2 (g) ➔ 2H_2O(l)$
3 $CaCO_3(s) ➔ CaO(s) + CO_2(g)$

Chemical calculations

A 35.5
A $2 \times 16 = 32$
A relative formula mass of $CaCO_3 = 40 + 12 + (3 \times 16) = 100$
  $40 \times 100 = 40\%$
1 24
2 $12 + (4 \times 1) = 16$
3 relative formula mass $= 14 + (3 \times 1) = 17$
  % of hydrogen $= 3 \div 17 \times 100 = 17.6\%$

Acids and their reactions

A pH 1 or 2
A calcium sulphate
A hydrogen
1 hydrogen ion (H^+)
2 copper chloride, carbon dioxide and water
3 nitric; hydrogen

Bases and neutralisation

A alkali is a soluble base
A potassium oxide
A water
1 hydroxide ion (OH^-)
2 an alkali
3 sodium chloride + water

Rates of reaction I

A mass will decrease
A the steepness increases, the faster the rate
A substance which speeds up a reaction without taking part in the reaction
1 because it lowers the activation energy, which means at lower temperature there are still successful collisions
2 faster, substance, faster, temperature

Rates of reaction II

A time
A exothermic
A the reaction is reversible
1 a) oxygen + water; b) enzyme/catalyst;
c)

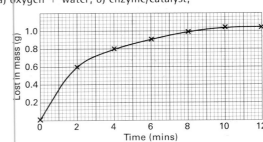

d) because as time goes on there are less particles to react

Enzymes

A amylase
A lime-water turns cloudy in the presence of carbon dioxide
A lipids or fats
A heated to a high enough temperature to kill bacteria in milk
1 enzyme that uses sugar as food
2 protease
3 reduce the temperature (therefore cost) of reactions

Useful products from oil

A   yes it is
A   cracking breaks up a compound using heat
A   at the top of the column
1   carbon and hydrogen
2   $CH_4 + 2O_2 \rightarrow CO_2 + 2H_2O$

Ammonia and fertilisers

A   nitrogen
A   contains nitrogen, which improves the growth of plants
A   too much nitrate in drinking water
1   three different elements: nitrogen, hydrogen and oxygen
2   (i) dissolves and is washed into rivers
    (ii) it is soluble in water
    (iii) causes them to grow and multiply

Extraction of metals

A   negative (the aluminium ions are positive)
A   it reacts with the impurities to produce slag
A   negative electrode
1   electrolysis as it is high in the reactivity series.
2   carbon dioxide
3   copper sulphate solution

Rocks and the rock cycle

A   igneous rocks formed inside the Earth's crust
A   removal of land by weathering
A   marble
A   E
1   a) A: metamorphic rocks B: igneous rocks; b) sediments are compressed; c) plates colliding
2   sedimentary rocks

Changes in the atmosphere

A   condensation of water vapour
A   carbon dioxide
A   via respiration and decay
1   the plants evolved and produced oxygen as a product of photosynthesis
2   a) respiration, burning, decay by bacteria; b) photosynthesis, which reduces the amount of carbon dioxide in the air; c) increase the amount of carbon dioxide in the air, global warming

Electric current and charge

A   the ammeter is placed in series so that it can measure the rate of flow of charge through the circuit
A   cell, lamp, ammeter, resistor
1   electric charge; equal to
2   10 minutes = 10 × 60 seconds = 600 seconds
    Q = It = 5 A × 600 seconds = 3000 C

Voltage and resistance

A   Q = It = 180 A × 6 × 60 seconds = 64800 C
A   V = I R = 2 A × 10 Ω = 20 V
1   energy; smaller; larger; brighter

Electric current in circuits

A   V = IR
A   The resistance of a wire increases as it gets hotter.
A   Because it only allows current to pass in one direction.
1   increases

More about electric current in circuits

A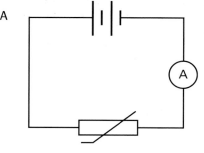

A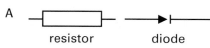
    resistor   diode

    LDR   thermistor

1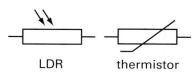

2   increases; decreases.

Electricity, energy and power

A   6 kW
A   a cell or battery
A   2 kW × 3 h = 6 kWh
A   1.5 ÷ 3 × 100% = 50%
1   P = V I = 12 V × 3 A = 36 W
2   energy = power × time = 12 W × (10 × 60) s
    = 12 W × 600 s = 7200 J
3   6 ÷ 15 × 100% = 40%

Electricity at home

A   direct current
A   If the fuse rating is much higher than the current needed by the appliance a fault may cause too much current to flow before the fuse breaks the circuit.
1   conductor: copper; insulator: plastic; low melting point: thin resistance wire
2   13 A fuse needed. The current through the kettle is about 9A.
3   Direct current always flows in the same direction. Alternating current changes direction and size regularly.

Electric charge

A   Because all the hairs have the same electric charge and repel each other, getting as far apart as possible. The crackle happens when the charges discharge, making small sparks.

A   The TV picture is made by an electron beam inside the television tube. The electrons bombarding the screen cause a charge to build up on the glass. This charge attracts charged particles of dust, which stick to the glass.

1   As petrol passes along the rubber hose the motion can cause static charge to build up. The wire allows the charge to leak away before it builds up and causes a spark.

2   When Julie walks across the carpet she becomes charged. The metal rail is a good conductor, so when she touches it the charge from her hand jumps across the gap, giving her a shock.

Electromagnetism

A   The magnetic field, due to an electric current in a coil, can be made stronger by increasing the current, increasing the number of turns in the same length and by putting iron through the coil.

A   By increasing the current in the wire or by increasing the strength of the magnetic field.

A   More current through the coil, more turns of wire on the coil or a stronger magnetic field.

1   circular; field; force; current; field

2
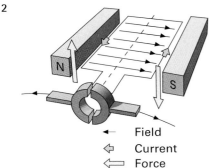

←   Field

⇐   Current

⇐   Force

Electromagnetic induction

A   A magnet to move into a coil or a wire to move through a magnetic field, induces a voltage.

A   The current would be induced in the opposite direction.

A   An alternating current in the primary makes a changing magnetic field which is needed to induce a voltage in the secondary coil.

1   (a) flick to left; (b) flick to left (c) flick to left; (d) needle stationary

Distance, speed and acceleration

A   distance = speed × time = 30 m/s × 20 s = 600 m

A   Velocity tells us the direction something is moving as well as the rate at which it is moving. Speed only tells us the rate at which it is moving.

A   17.5 m/s^2 is nearly twice the acceleration due to gravity. You would feel very heavy, pressing down into your seat as the shuttle lifts off.

1   speed = distance ÷ time = 400 ÷ 8 = 5 m/s

2   0–A accelerating from rest to 10 m/s over 20 s; A–B steady speed of 10 m/s for 40 s; B–C decelerates from 10 m/s to rest in 20 s; C–D stationary for 20 s; D–E accelerates to 7 m/s over 10 s; E–F steady speed of 7 m/s for 30 s; F–G decelerates to rest over 15 s.

3   acceleration = change in velocity ÷ time taken
    = 30 ÷ 10 = 3 m/s^2

Forces – in pairs and balanced

A   The forces on an object moving at a steady speed are balanced – or there are no forces acting.

A   Gravity acts down on the book. The table pushes up with a reaction force.

1   Parachutist : gravity and drag; cyclist: muscles, gravity and drag; car: engine and drag

2   acting

Forces – mass and acceleration

A   The forces acting on an object that is accelerating are unbalanced. The net force is in the direction in which it is accelerating.

A   The bigger the mass, the smaller the acceleration.

1   the cyclist is slowing down, the unbalanced force is backwards; car is accelerating – unbalanced force is forwards; parachutist is falling at a constant speed – the forces are equal and opposite, so are balanced; the weights are stationary as the forces on it are balanced.

Stopping forces

A   There are resistive forces between the moving parts in a car. There are resistive forces between the tyres and the road – that enable it to move. There is also air resistance as the car moves through the air.

A   Tiredness, drugs, alcohol, distractions inside the car, poor visibility.

1   The kinetic energy of the moving object causes the rise in temperature.

2   If the driver is tired his reactions will be slow, it will take longer for him to apply the brakes and the car will have travelled further before it stops.

Stopping a car

A   mass of car, road conditions, car brakes

A   Time it takes for the driver to react and the distance the car travels when braking.

1   (a) A heavily-loaded car has a large mass so the deceleration will be less and it will travel further while braking- the stopping distance will be greater.
    (b) The car is moving quickly, it has a greater change in speed to make so it will need to decelerate for longer and travel further – a greater braking distance. It will also travel further during the drivers reaction time, so the thinking distance will be greater. The stopping distance will be greater.
    (c) The friction between the tyres and a wet road will be less, the driver will have to apply the brakes gently to avoid skidding so the car will travel further while braking and the stopping distance will be longer the mass of the car; the speed the car was travelling before braking; the friction forces in the brakes; the road conditions

2   (a) distance = speed × time = 30 m/s × 0.5 s = 15 m
    (b) the thinking distance would be longer – the driver's reaction time would be longer.
    (c) total stopping distance = thinking distance + braking distance = 15 m + 64 m = 79 m

Falling

A You need to know your mass in kilograms. The force of gravity on you (your weight) is 10 newtons for each kilogram.

A Air resistance depends on the shape of the object and the speed at which it falls.

A The terminal velocity of an object depends on the weight of the object and its shape.

1 (a) The force down on the parachutist is always the same – her weight. When she lies across the air her air resistance is quite high, when she become vertical her air resistance decreases, until the parachute is fully open when the air resistance becomes equal to her weight.
(b) Initially the parachutist accelerates, but the acceleration decreases as the air resistance increases, until he air resistance is equal to her weight, when her speed is a maximum – terminal speed.

2 The aircraft needs to stop over a much shorter distance, so needs a large stopping force.

Properties of waves

A The cork floating on the water just bobs up and down but does not move along with the wave.

A angle of incidence = angle of reflection

A Refraction occurs because waves change speed when they pass from one material to another.

A Light passing from glass to air increases in speed and is refracted away from the normal.

1

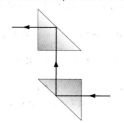

Electromagnetic spectrum

A Gamma rays carry more energy so can cause more damage to cells in the body.

A X-rays can damage cells in the body by ionising molecules in the cell. The greater the exposure to x-rays, the greater the chances of lasting damage.

1 gamma rays ➔ x-rays ➔ ultraviolet ➔ visible ➔ infrared ➔ microwaves ➔ radio waves

2 (a) x-rays used for imaging – a hazard because they damage cells in the body
(b) microwaves used in communications and cooking; hazardous because they some wavelengths cause heating of water.
(c) ultraviolet waves used to sterilise surgical equipment, hazardous because they can cause skin cancer.

Communicating with waves

A Optical fibres can carry more information than an equivalent thickness of copper. They cannot easily be 'tapped'.

A Microwaves travel in straight lines – the Earth curves.

A Diffraction happens best when the gap the wave passes through is about the same size as the wavelength of the wave.

A Digital signals can carry more information and suffer less from interference.

1 Optical fibres can carry more signals than copper wire; the fibres are cheaper than copper; the signal does not escape from the cable, so cannot be detected; the energy loss in the fibre is less than in copper; the signal suffers less from interference.

2 Instead of a sharp shadow, the edges of the shadow blur or even show fringes of light beyond the edge of the shadow.

3 Analogue signals can have any value, related to the size of the original sound. A digital signal is a string of 0s and 1s, which are binary numbers, to show how the voltage of the signal is changing.

Sound and ultrasound

A Sound waves are longitudinal waves

A X-rays damage cells in the body – there are no known side-effects to using ultrasound.

1 (a) frequency = vibrations ÷ time = $100 \div 2 = 50$ Hz
(b) Shorten the length vibrating to make a higher pitched note.

2 Sound travels 60 m (there and back) in 0.2 s.

3 You can hear round corners – you do not need to be in direct line of sight with a sound source to hear it.

Structure of the Earth

A Plates move apart, allowing molten rock to come to the surface. Plates move together, forming mountains by folding; causing volcanoes to erupt, or earthquakes. Plates slide past each other causing earthquakes.

A The Atlantic ridge is formed at the junction between the South American and North American plates and the African and Eurasian plates.

A Primary (P) waves and Secondary (S) waves.

1 P waves are longitudinal waves; S waves are transverse and travel more slowly.

2 P waves arrive first.

The solar system

A Mercury, Venus, Earth, Mars, Jupiter, Saturn, Uranus, Neptune, Pluto

A 7

1 Between Saturn and Uranus

2 The further a planet is from the Sun, the longer it takes to complete one orbit.

Gravity

A Gravity is less on the Moon than the Earth because the mass of the Moon is much less than the mass of the Earth.

A Telephone transmissions, television transmissions, weather monitoring, environmental monitoring, spying, astronomical observations.

1 The asteroid is very small compared to the Earth, so the mass will be small and its gravitational pull will be small – it will be much easier to leave the surface.

Our place in the Universe

A It is expanding

A Only you can answer this – you might be happy to accept the evidence of radio signals that seemed to make a pattern to communicate – or you might want to see the living things with your own eyes – or something in between.

1 Our Sun will probably become a white dwarf, gradually fading as it cools down.

2 distance = speed × time = 300 000 km/s × 480 s
= 144 million km

A

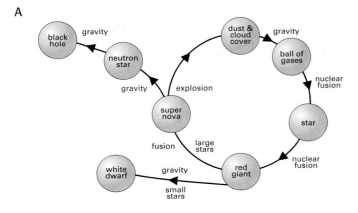

Work, power and energy

A work done = force × distance = 500 N × 2 m = 1000 J
A from the food he eats
A potential energy = m g h
   = 70 kg × 10 N/kg × 3 m = 2100 J
A The energy shakes the molecules in the ground and the cans so they get a little warmer; some energy is carried away by sound.
1 The aeroplane has kinetic energy and gravitational potential energy gained from the fuel burned in the engines.
   The book on the shelf has gravitational potential energy – gained from the muscles of the person who put it there.
   The cyclist and bike have kinetic energy gained from the food eaten by the cyclist and transferred through muscles.

Energy transfers

A saucepan is a good conductor (handle should be a poor conductor), jacket is a poor conductor
A Convection currents carry the warm air up the staircase.
A When the sweat evaporates it carries away energy from the surface of your skin.
A Frying in a pan uses conduction to transfer energy from the cooker to the food; cooking under the grill uses radiation; cooking soup in a saucepan uses convection for the energy from the bottom of the pan to be transferred through the soup.
1 fibre glass, loose fill polystyrene
2 Energy is transferred round the room by convection currents rather than by radiation – the radiators are often painted a light colour, which makes them poor radiators.
3 Energy is lost through the roof by convection and conduction. A layer of insulating material creates air pockets, which are poor conductors and prevent convection.
   Energy is lost through drafts through gaps between doors and windows and their frames. Draught proofing materials can be stuck to the frame to make a better seal.
   Energy is lost by conduction through the glass of windows; double-glazing inserts an insulating layer of air between to layers of glass.

Energy resources

A All the energy eventually warms up the surroundings – through friction, braking and air resistance
A Generating electricity by fossil fuels causes increase in acid rain and global warming; using fossil fuels for electricity generation means they are not available for other uses – such as making materials; the fossil fuel supply is limited and will run out

A The temperature has to drop lower before the heating switches on – so the heating is on for less time.
1 Fossil fuels are used to make plastics
2 Only 35% of the energy from the coal becomes a useful output as electricity.
3 A bus carries 30–80 people, so reducing the number of cars on the road, less fuel is used.
4 Advantages of wind and water are that they are renewable resources that will not run out and they are non polluting, and do not increase global warming. Disadvantage of wind power is that it is only available when the wind blows; water power depends on a good water supply, so depends on being in the right place.

What is radioactivity?

A Protons and neutrons.
A 2 protons and 2 neutrons.
A Negative charge.
A Lead stops gamma radiation.
1 alpha α: helium nucleus – 2 protons and 2 neutrons; few centimetres; paper beta β: electron; few metres; few millimetres of aluminium gamma γ: electromagnetic radiation very long way; thick piece of lead or thicker concrete
2 Beta particles.
3 Alpha, beta, gamma.

Radioactive decay

A Do not eat seafood, move to a non-granite part of the country, don't fly in aeroplanes.
A Half-life is the time for half the radioactive nuclei to decay
1 (a) six days is three half lives
   (b) The amount will have halved three times: 12 mg 96 mg 93 mg 91.5 mg
2 It loses three quarters which means one quarter is left. In 1 half life it falls to half, in second half life it falls to one quarter – so two half lives, 110 s.

Radioisotopes and carbon dating

A Radiocarbon dating is for things made from once-living material.
1 about 750 million years
2 The test was carried out by three independent groups to make sure there was no contamination and to increase the accuracy of the result.

Using radioactivity

A The fact that radiation ionises is used in detection, its properties of penetration help identify the type of radiation.
A Beta radiation would lose most of its energy before reaching the cancer, and cause possible damage on the way.
A Gamma penetrates paper easily and there would be no detectable change in intensity after passing through paper.
1 Alpha emitters have such a low penetration, they would damage the cells close to the source, rather than the cancerous cells.
2 Gamma is used because it can penetrate the ground and so be detected at the surface.

A

acceleration The rate at which a moving object is speeding up or slowing down. Measured in metres / second² (m/s²).

acid A solution with a pH of less than 7, which contains hydrogen ions.

ADH Antidiuretic hormone which controls the production of urine in the kidneys.

aerobic respiration Oxygen is used in the process of transferring energy from glucose in the cells of the body.

alkali metals Group I of the periodic table. They have one electron in the outer shell.

alkalis Form hydroxide ions when added to water. Alkalis have a pH greater than 7.

alkenes Unsaturated hydrocarbons with a double bond between the carbon atoms.

allele One version of the gene coding for a particular characteristic.

alternating current (a.c.) An electric current which regularly changes size and direction.

amp Electric current is measured in amps. 1 amp = 1 coulomb / second.

amplitude Distance from the crest of a wave to the place where there is no displacement.

anaerobic respiration Energy is transferred from glucose without oxygen.

analogue signal Carries information by copying the changing pattern of the waves in the original information.

angle of incidence The angle between the direction of a wave and the normal.

artery A blood vessel with a thick muscular wall and a narrow space inside.

atom All elements are made of atoms. Consists of a nucleus containing protons and neutrons surrounded by electrons.

atomic number (Z) The number of protons in the nucleus of the atom.

auxins Plant hormones found in the tips of growing shoots and roots. The hormones control the growth of the plant.

B

background radiation We are surrounded by radioactive materials, in rocks, in the air we breathe and particles from outer space. All these sources of radioactive particles are called background radiation.

base Metal hydroxides, oxides, carbonates. Bases that dissolve in water are alkalis.

C

capillary A thin-walled, narrow blood vessel.

carbohydrates Foods made of carbon, hydrogen and oxygen; supply energy.

carbon cycle Shows how the carbon that all living things need for growth is recycled via the carbon dioxide in the atmosphere.

carnivore Animal that eats other animals.

catalyst Changes the rate of a chemical reaction without being changed itself.

cell All living things are made of cells; the smallest units of living matter. All cells have a nucleus, cytoplasm and surface membrane.

chlorophyll Causes the green colour in plants, necessary for photosynthesis.

chloroplast In plant cells, they contain chlorophyll for photosynthesis.

chromosome The cell nucleus contains chromosome pairs which contain genes.

cloning A process of making an identical copy of a living thing.

combustion Reaction of a substance burned in oxygen. An oxidation reaction.

compound Elements join together to form compounds by forming bonds.

concentration How much substance is dissolved in water. Measured in moles per litre or moles per dm³. Higher concentration: more particles of the substance present.

condensation Molecules in a vapour return to a liquid, losing energy as they do.

conductor A material that allows an electrical current to pass easily. It has a low resistance. Allows thermal energy to be transferred through it easily.

cornea Transparent outer coating of eye.

coulomb (C) Measures electric charge.

covalent bond Bond between atoms forms when atoms share electrons to achieve a full outer shell of electrons.

cracking A large molecule is broken into smaller molecules using a catalyst.

critical angle Angle of incidence that results in waves being refracted through 90°.

cytoplasm Living contents of the cell where the chemical processes take place.

D

decomposition Reaction in which substances are broken down, by heat, by electrolysis or by a catalyst.

diffraction When waves pass through a gap comparable in size to the wavelength.

diffusion Effect of randomly-moving particles in a liquid or gas gradually spreading from an area of high concentration to one of low concentration.

digestion Food passes through the gut and is broken down into small enough molecules to pass into the blood. Undigested food passes out of the anus as solid waste.

digital signal Carries information as 1s and 0s to code the values of the original signal.

diode Electrical component that only allows electric current to pass in one direction.

direct current (d.c.) Always flows in the same direction.

distillation Separation of a mixture by evaporation and condensation of the components with the lower boiling points.

DNA Chemical which makes chromosomes.

E

effector Part of the body which responds to a stimulus, such as a muscle or gland.

efficiency Measure of how effectively energy is transferred in a system.

electric charge Electrons carry a negative charge, protons carry a positive charge. Electric charges attract and repel each other. Electric charge is measured in coulombs (C).

electric current A flow of electric charge around a circuit, measured in amperes (A).

electrolysis When an electric current passes through a solution or molten solid.

electrolyte Liquid which conducts an electric current during electrolysis. It contains ions which carry the current.

electromagnetic induction When a conducting wire moves relative to a magnetic field, a voltage is induced across the wire.

electromagnetic spectrum Family of waves which travel at the same speed in a vacuum.

electron Very small, negatively-charged particle surrounding the nucleus in an atom. No. of electrons around nucleus = no. of protons in nucleus, in a neutral atom.

electronic configuration Arrangement of electrons in energy levels around the nucleus.

element All atoms of an element have the same atomic number, the same number of protons and electrons and so the same chemical properties.

enzymes Made of protein molecules. They are biological catalysts, molecules become temporarily attached to the enzyme during biological processes.

epicentre The place in the Earth that an earthquake occurred. Seismic waves spread out from the epicentre.

equilibrium When the rate of the forward reaction equals the rate of the back reaction in a reversible reaction.

evaporation Molecules near the surface of a liquid may leave the liquid as a vapour.

exhalation (expiration) Process of decreasing the space inside the chest so air pressure rises and air flows out of the lungs.

fertilisers Substances containing nitrogen, which are used for healthy growth of plants.

food chain Shows the feeding relationship between living things and the way energy and nutrients flow through the system.

food pyramid Illustrates the numbers of the different living organisms in a food chain.

food web Shows how food chains are interlinked and how organisms in an ecosystem depend on each other.

fossil fuel Created over millions of years by the decay and compression of living things, particularly plants.

fractional distillation Mixture of several substances (e.g. crude oil) are distilled. The evaporated components are collected as they condense at different temperatures.

frequency Number of waves produced each second, in Hertz (Hz).

fusion Two atomic nuclei collide with such energy that they fuse, releasing energy as electromagnetic waves.

galaxy Collection of many stars held together by gravity.

galvanising Protecting a metal from corroding by coating it with zinc.

gene Short length of DNA which is the code for a protein. There are many genes within each chromosome.

genetic engineering Involves taking genes from one living thing and inserting them into the DNA of another to change its characteristics.

giant lattice Ionic compounds form a giant lattice structure.

giant structure Some covalent bonds produce very strong, giant structures.

glucose Simple carbohydrate produced by plants during photosynthesis. Some is used as an energy source by the plant, the rest is converted to starch and stored in the leaves.

gravitational potential energy When an object is lifted against the pull of gravity, energy transfers to the gravitational field.

gravity Force that pulls objects to the ground. Gravitational attraction acts between any two objects. Strength of the gravitational field is measured in newtons / kilogram (N/kg).

groups Elements in the periodic table with the same number of electrons in their outer shells and so have similar chemical properties. A group of elements lie in the same column.

half-life The time it takes for half a quantity of radioactive material to decay to a new substance.

halogens Elements in group VII with 7 electrons in the outer shell.

herbivore Animal that only eats plants.

hertz Measurement of frequency.

hormone chemical which helps to co-ordinate life processes.

hydrocarbons Group of compounds containing only hydrogen and carbon. Many are derived from oil.

igneous rock Forms when molten rock cools.

inhalation (inspiration) The process of increasing the space inside the chest so that the air pressure drops and air flows in to the lungs.

inherited disease Disease passed on from one generation to the next from the genes of the parents to the child.

insulator A material which does not allow an electrical current to pass.

insulin Hormone involved in the control of sugar levels in blood.

ion Atoms can lose or gain electrons. On losing an electron they form a positive ion. On gaining an electron they form a negative ion.

ionic bond Forms when an electron is transferred from one atom to another, forming a positive-negative ion pair.

iris Opaque tissue, which controls the size of pupil, controlling the amount of light which can pass into the eye.

isotope Atoms of the same element with different numbers of neutrons

joule Energy is measured in joules.

kidney Organ that controls water and salt balance and the pH of the blood. Kidneys excrete urea, hormones and medicines in the urine they produce.

kinetic energy Energy transferred to a moving object when it is accelerated.

LDR (light-dependent resistor) Resistance value of the component depends on the level of light falling on the component.

longitudinal wave Wave whose oscillations travel in the same direction as the energy.

magnetic field The region in space around a magnet in which other magnets are affected.

mass Amount of material in an object in kg.

mass number (A) Number of protons and neutrons in the nucleus of an atom.

meiosis Parent cell divides to create four new sex cells, each with a different combination of chromosomes.

metal Elements that are shiny, good conductors of heat and electricity.

metamorphic rocks Formed when rocks are changed by heat or pressure.

microbe Single-cell organisms, e.g. bacteria.

mitosis Process of cell division which results in two identical cells.

mole Relative molecular/formula mass of a substance in grams.

monomer A simple molecule.

motor neurone Nerve cell that transmits a signal from the spinal cord to the effector.

mutation Spontaneous change in a gene or chromosome, which may result in a change in cells characterised by the gene.

negative ion (anion) When an atom gains an electron to fill the outer shell of electrons.

nephron Structure within the kidney which filters out waste from the blood and allows useful materials to be reabsorbed.

neutralisation Reaction between an acid and a base, producing salts and water.

neutron Small, uncharged particle, with the same mass as a proton, found in the nucleus of the atom.

newton (N) Measurement of force. 1 newton is the unbalanced force needed to give a mass of 1 kg an acceleration of 1 m/s^2.

nitrogen cycle All living things need nitrogen to make proteins. The nitrogen cycle shows how nitrogen is recycled so it is available to both plants and animals.

noble gases Elements in group 0/VIII of the periodic table, with a full outer shell of electrons and so are unreactive.

normal Line perpendicular to the surface at the point where a wave crosses the surface.

nucleus of a cell Made of genetic material, such as DNA. The nucleus contains the instructions for the cell.

nucleus of an atom An atom has a small, dense, positively-charged nucleus where most of the mass is concentrated, containing protons and neutrons.

oestrogen One of the sex hormones in females which causes puberty.

ohm Measurement of electrical resistance.

optic nerve Nerve fibres from the light-sensitive cells on the retina form the optic nerve, passing from the eye to the brain.

optical fibres Very fine threads of glass through which light can pass, even when the fibre is curved around a corner.

ore Rock containing a mineral which can be extracted from the rock.

osmosis A special case of diffusion – a solvent passes through a membrane from a weak to a stronger solution. The solute cannot pass through the membrane.

oxidation Reaction in which oxygen combines with a substance; a reaction where electrons are removed; a reaction where hydrogen is removed from a substance.

oxygen debt During vigorous exercise anaerobic respiration takes place in muscle tissue. Lactate is produced in the absence of oxygen. When exercise stops, oxygen is used to respire the lactate.

periodic table Elements in order of atomic number. The table is arranged into rows called periods and columns called groups.

periods Elements in which the same outer shell is being filled up.

pH Measure of how acid or alkali a solution is. The scale runs from 1 to 14.

photosynthesis Plants process water and carbon dioxide to produce glucose by the process of photosynthesis.

plasma Yellow liquid part of blood which contains many dissolved substances.

polymer Large molecule formed from many monomers by polymerisation.

polymerisation When many identical monomers join together forming a polymer.

positive ion (cation) When an atom loses an electron to form a full outer shell of electrons.

power Rate at which energy is transferred in a system. Power is measured in watts (W) and kilowatts (kW).

progesterone One of the sex hormones which causes puberty in females.

proteins Foods made from carbon, hydrogen and nitrogen. Proteins are used to build enzymes and cytoplasm.

proton Small, positive particle found in the nucleus of the atom.

pupil Light passes through this hole into eye.

radioactive decay Random process in which the nucleus of an atom becomes more stable by losing particles and energy.

reactivity series List of metals in order of decreasing reactivity. A metal will displace one below it in the series from its solution.

receptor Detects a stimulus.

red blood cells Contain haemoglobin which transports oxygen to cells and returns carbon dioxide to the lungs.

redox reaction Oxidation and reduction always take place together. The combined reaction is called a redox reaction.

reduction Reaction in which oxygen is removed from a substance; or a reaction where electrons are gained or a reaction where hydrogen is gained by a substance.

reflection Reflection of waves happens when the waves bounce off a surface.

refraction Refraction happens when waves change direction due to a change in speed.

relative atomic mass The relative mass of an atom of an element.

resistance Measure of difficulty for an electric current to pass. Measured in ohms (Ω).

respiration Process which transfers energy from food to the organism.

retina Light-sensitive layer at back of eye.

reversible reaction Chemical reactions that can go both ways. Direction of the reaction depends on the condition of the reactants.

rust Produced when iron corrodes, reacting with water and oxygen.

satellite Body in orbit around a planet. Moons are natural satellites. There are many man-made satellites in orbit around the Earth.

saturated compounds Atoms are joined together by single bonds.

sedimentary rocks Formed when rock fragments are deposited and then pressed together.

seismic waves Carry energy away from an earthquake.

sensory neurone Nerve cell that carries a signal from the receptor.

shells of atoms Electrons are grouped within an atom in regions of space called shells or energy levels.

solar system The Sun, nine planets, the asteroid belt and a number of comets.

speed Rate at which something moves. Speed is measured in metres / second (m/s).

stomata Pores (openings) in the leaf surface, through which transpiration occurs.

temperature regulation The body controls its internal temperature by varying blood flow, sweating and shivering.

terminal velocity When an object moves through a fluid there is resistance to motion. The faster it travels, the greater the resistance. When the driving force is balanced by the resistive forces, the object is moving at a maximum speed (terminal velocity).

testosterone Sex hormone which causes puberty in males.

thermal decomposition Reaction where a substance is broken down by heat.

thermistor Electrical component whose resistance changes with temperature.

total internal reflection When the angle of incidence is greater than the critical angle.

transformer Pair of wire coils linked by a piece of iron. A changing current in one coil creates a changing magnetic field in the iron, causing a changing voltage in the other coil by electromagnetic induction.

transition metals Group of elements in the periodic table between Group II and III.

transpiration Loss of water from a plant by evaporation from stomata.

transverse wave Oscillations at right angles to the direction in which the energy travels.

unsaturated compounds One or more bonds between atoms are double bonds.

vein Blood vessel with a thinner wall, larger space and lower pressure than an artery.

velocity Speed of a moving object and the direction it is moving.

villi The surface of the small intestine is lined with villi, small 'fingers' of gut which increase the surface area for digestion.

volt Measurement of electrical potential difference. 1 volt = 1 joule / coulomb.

voltage Potential difference between two points in a circuit measures the difference in energy carried by the electric charge passing through the circuit. Measured in volts (V).

watt Power is measured in watts (W). 1 watt is 1 joule / second.

wavelength Distance from the crest of one wave to the crest of the next wave.

weight Force of gravity on the object. Weight is measured in newtons (N).

white blood cells Fight infection – they make antibodies and overcome bacteria.

work Work is done when a force moves an object and energy is transferred. Work is measured in joules. When a force of 1 newton acts through 1 metre in the direction of the force, 1 joule of work is done.

Last-minute learner

- These six pages give you the most important facts in science in the smallest possible space.
- You can use these pages as a final check.
- You can also use them as you revise as a way to check your learning.
- Cut them out for quick and easy reference.

Cells

- Life processes are the main activities carried out by body systems: **respiration, feeding, sensitivity, movement, reproduction, growth, excretion**.
- The smallest unit of life is a cell. Groups of cells of the same type form tissues and tissues build organs which have specialised functions.
- Common cell structures include:
  the **cell surface membrane**, which controls what moves in and out of a cell; the **cytoplasm**, which is where chemical reactions happen, and the **nucleus**, which contains genetic material.
- Plant cells have additional structures: **cell vacuole**, which gives plants support and maintains shape; **chloroplasts** containing chlorophyll, which is important for photosynthesis and **cell walls**, which give support.

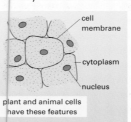

plant and animal cells have these features

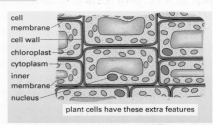

plant cells have these extra features

Life processes

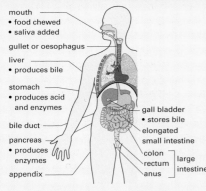

- mouth
  - food chewed
  - saliva added
- gullet or oesophagus
- liver
  - produces bile
- stomach
  - produces acid and enzymes
- bile duct
- pancreas
  - produces enzymes
- appendix
- gall bladder
  - stores bile
- elongated small intestine
- colon
- rectum | large intestine
- anus

- We get nutrition/nutrients from food, providing **raw materials** to build new cells and **energy** for life processes.
- A balanced diet includes: **carbohydrates, proteins, fats, water, minerals** and **vitamins** in the correct proportions and in levels appropriate to lifestyle.
- Food is broken down into smaller particles as it passes through the gut. Digested particles are absorbed into the bloodstream for transport to body tissues.
- Enzymes are mainly responsible for digestion: **carbohydrases** digest carbohydrates such as starch; **proteases** digest proteins; **lipases** digest fats/lipids.
- The shape of an enzyme molecule is vital to its function. The active site is a part of the enzyme where molecules 'lock on' and react. Changes in pH, temperature or other conditions can destroy an enzyme by changing its shape.
- The heart is muscular and beats throughout life. At each beat the atria fill and contract, and then the ventricles fill and contract. The left and right sides beat simultaneously but the blood flow is separate.
- Arteries, veins and capillaries are the blood vessels that form the circulation to all cells. Blood contains dissolved substances, e.g. glucose and salts, plasma proteins (e.g. antibodies), platelets and red and white cells.
- Ill-health may be caused by injury, inherited disease, or infection. Micro-organisms may cause infection.
- The body's main defences against infection include skin, cilia and white blood cells.
- Immunisation is a way of protecting ourselves against disease by causing an immune response.
- Lifestyle (diet, exercise and drug abuse, including smoking and drinking alcohol) has a significant effect on health.
- Breathing movements change the air pressure inside the lungs. During inhalation the pressure inside the lungs is less than ouside the body, and vice versa in exhalation.
- Respiration transfers energy from food to living cells. Aerobic respiration involves oxygen:

glucose + oxygen → carbon dioxide + water
(+ energy transferred)

Anaerobic respiration happens in the absence of oxygen:

glucose → lactate (+ some energy transferred)

glucose → ethanol + carbon dioxide
(+ some energy transferred)

Variation, genetics and ecology

- **Variation** describes the differences between living things. It is caused **genetically** through inherited genes, when chromosomes exchange DNA during cell division; or mutation can occur; or by **environmental factors**.
- A **gene** is a short chunk of DNA, which codes for a particular protein giving rise to a characteristic.
- **Charles Darwin** and **Alfred Wallace** contributed to the theory of evolution by natural selection: **mutations cause genetic variation**, and give rise to new characteristics; a new characteristic may be **beneficial or harmful to the survival** of an individual; a beneficial characteristic is **more likely to be passed on** to future generations.
- **Sexual reproduction** involves two parents, each providing a nucleus from a sex cell. There is genetic variation in offspring. **Selective breeding** involves breeding plants or animals with desirable characteristics, improving them over several generations.
- **Asexual reproduction** uses one parent to produce genetically identical offspring that are identical to the parent. Cloning is a form of asexual reproduction used in horticulture (e.g. plant cuttings) and agriculture (e.g. splitting embryos in cattle breeding).
- **Genetic engineering** transfers genes artificially from one organism to another e.g. human insulin gene to bacteria.
- A gene codes for a particular characteristic. Alleles are different forms of the same gene. An offspring inherits two alleles for a characteristic, one from each parent.
- A **dominant allele** will hide a recessive allele's appearance, even if there is only one. Write a dominant allele with a capital letter and a recessive allele with a lower case letter.
- To complete a **crossing diagram** write the alleles of the parents alongside the crossing diagram, fill in the possible combinations for each square and decide how each offspring will look according to alleles inherited.
- An **ecosystem** is made up of living things, the environment and all their interactions. Each ecosystem has its own variety of living things (biodiversity), which is reduced by the impact of human activities.
- There are **competing priorities** for use of the Earth's resources because people have different priorities.
- Strategies for protecting the future of our planet include: recycling materials, conservation programmes, protecting sensitive ecosystems, programmes aimed at protecting endangered species, managing ecosystems.
- Living things are linked by **food chains**. Plants are mostly at the start of a chain. Animals are consumers, because they eat ready-made food in the form of plants or other animals. Many food chains overlap, forming a food web.
- **Predator/prey populations** are interdependent: after a time lag the predator population follows that of the prey.
- Carbon dioxide is removed from the atmosphere by plants, during photosynthesis. It is added to the atmosphere by living things when they carry out respiration, when they decay or when we burn fossil fuels or other materials from living things e.g. wood.

Regulating body processes

- Hormones are chemicals that help to control conditions within the body and coordinate life processes. They are carried in blood and act on target organs.

part of the nervous system	what it does
receptor cells eg skin, organ	detects a stimulus, such as a change in temperature
sensory neurone/nerve	carries impulses from receptors to central nervous system (CNS)
central nervous system (brain and spinal cord)	processes information detected by receptors/sense organs
motor neurone/nerve	carries impulses from CNS to effector, e.g. muscle or gland
effector tissue or organ	carry out action eg gland makes hormone, muscle moves

- Skin helps regulate body temperature by adjusting the blood circulation near the body surface and the amount of sweat produced; it is waterproof and protects the body from drying out and from infection

- The menstrual cycle is 28 days. Ovulation occurs on day 14; fertilisation happens in the oviduct/Fallopian tube; implantation happens in the uterus.

gland	hormone and result
pituitary	important in controlling growth rate; menstrual cycle; milk production; the thyroid
thyroid	thyroxine: regulates rate of chemical activities
pancreas	insulin: reduces blood sugar level glucagon: increases blood sugar level
adrenal	adrenaline: prepares body for rapid activity
testis	testosterone: causes secondary sexual characteristics and sperm production
ovary	oestrogen: causes secondary sexual characteristics and helps control menstruation

All about plants

- **Photosynthesis** takes place in the choroplasts, where light energy is absorbed by chlorophyll. Simple raw materials are converted into carbohydrate and oxygen.

$$\text{carbon dioxide + water} \xrightarrow{\text{sunlight}} \text{glucose + oxygen}$$

- Plants use glucose as a raw material for making other substances and for energy.
- The **rate of photosynthesis** is determined by the conditions: light intensity, concentration of carbon dioxide and a suitable temperature.
- **Osmosis** is the diffusion of water molecules, which follows a water concentration gradient.
- **Transpiration** is the loss of water from plant surfaces, which draws water up through the plant. Transpiration happens quickest in warm, dry and windy conditions.
- Hormones are chemicals that coordinate plant growth by influencing how fast cells divide, and cell elongation.
- **Auxin** is a plant hormone, which causes uneven growth on either side of a shoot or root tip. In shoots it increases the rate of cell growth. In roots it slows the rate of cell growth.
- The growth of a plant towards light is **phototropism**. There is more auxin on the side of the shoot away from light, so it grows faster and curves towards light.

The periodic table

- The periodic table contains all known elements in **order of atomic number**.
- **Groups** go down the table and depend on the number of electrons the element has in its last shell.
- **Periods** go across the table, depending on the outer electron shell that is filling up.

- Metals are found on the left-hand side and non-metals on the right hand side.
- The main properties of metals are: **good conductors** of heat and electricity, and have **high melting points**.
- An alloy is a mixture of metals.

Group 1 metals

- E.g: lithium, sodium and potassium.
- These are the **most reactive metals** in the periodic table.
- They react with water to form metal hydroxides and H_2.
- They react with halogens to form metal halides.
- Group I compounds are ionic, white and soluble in water.

Transition metals

- The transition metals are found between groups II and III in the periodic table.
- They are 'typical' metals. They have high melting points and high densities.
- They are ductile and malleable.
- Transition metal compounds can often be coloured.
- Iron is a transition metal. Iron rusts when it reacts with air and water.

Reactivity series

- If a metal is higher in the series than the metal in a compound, displacement will take place.
- The reactivity series is determined by how vigorously the metal reacts with oxygen, water and acid.
- Metals react with oxygen in the air to produce a metal oxide.
- Metals react with water to produce metal hydroxide and hydrogen.
- Metals react with acid to produce a salt and hydrogen.

The non-metals

- The halogens are in group VII of the periodic table.
- They are **fluorine** (clear gas), **chlorine** (green-yellow gas), **bromine** (orange liquid) and **iodine** (purple solid).
- The **noble gases** are in group VIII (also called group 0).
- They are **unreactive gases** as they have full outer shells of electrons.

Ionic and covalent bonding

Ionic bonding

- Ionic bonds are formed between metals and non-metals.
- They are formed by **transferring electrons** from the metal to the non-metal.
- Properties of ionic compounds include **high melting and boiling points** and **conducting electricity** when they are molten and in solution.

Covalent bonding

- Covalent bonds are formed between **non metals**.
- They are formed by elements **sharing electrons**.
- They form molecules that have low melting points and do not conduct electricity.

Atoms, elements and compounds

- **Elements** are made up of atoms.
- An atom contains **protons**, **neutrons** and **electrons**.
- **Atomic number** = number of protons (or electrons).
- **Mass number** = number of protons and neutrons.
- **Isotopes** of an element have the same number of protons but different number of neutrons.
- Electrons are arranged around the nucleus in shell 1. The first shell can hold 2 electrons, and the 2nd and 3rd can hold 8 electrons.

Acids and ions

- Acids have a **pH less than 7**.
- All acids contain a **positive hydrogen ion** (H^+).
- Acids react with most metals to form a salt and H_2.
- Acids react with metal carbonates to produce a salt, CO_2 and water.

Bases and neutralisation

- Bases are metal oxides and hydroxides.
- They have a **pH greater than 7**.
- Bases that **dissolve in water** are called alkalis.
- Acids react with bases to produce salt and water, this is called **neutralisation**.
- Neutralisation reactions can cure indigestion.

Rates of reaction

- Rates of reaction can be affected by **concentration**, **temperature**, **surface area** and adding a **catalyst**.
- The **collision theory** states that particles will react if they collide with sufficient energy.
- A reversible reaction is a reaction which can go both ways:

$$\text{products} \rightleftharpoons \text{reactants}$$

- An exothermic reaction gives out heat and an endothermic reaction takes in heat.

Enzymes

- Enzymes are **biological catalysts** – they speed up reactions.
- **Fermentation** uses the enzyme, yeast, to produce alcohol and carbon dioxide from sugar.
- Enzymes are used to make **yoghurt** and **bread**.
- In industry enzymes are used to stone-wash denim, in washing powders and in baby foods.

Industrial processes

Products from oil

- Crude oil is separated into its components by **fractional distillation**.
- Crude oil contains hydrocarbons.
- When hydrocarbons are combusted they produce carbon dioxide and water.
- **Cracking** is when long-chain hydrocarbons are broken into smaller hydrocarbons.

Ammonia and fertilisers

- Ammonia is made from **nitrogen** and **hydrogen** and has the formula NH_3.
- Fertilisers are used to **promote plant growth**. Two important fertilisers are **ammonium nitrate** and **ammonium sulphate**.
- If excess fertiliser gets into a river or stream it can kill the plant and animal life. This is called eutrophication

Extraction of metals

- **Aluminium** is extracted using electrolysis.
- **Iron** is extracted using the blast furnace. The reducing agent for this process is **carbon monoxide**.
- **Copper** is purified using electrolysis.

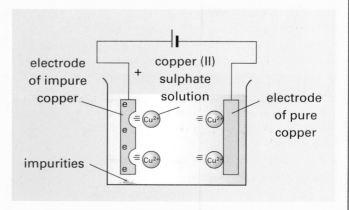

Changes in the atmosphere

- In the beginning, when the Earth's surface solidified, there were many volcanoes erupting, which gave off **carbon dioxide**, **ammonia** and **methane**.
- The water vapour condensed to form the oceans.
- There was little oxygen in the atmosphere.
- Plants started to give out oxygen and take in carbon dioxide 500 million years ago.
- The methane and ammonia reacted with the oxygen.
- The air today is 78% nitrogen, 21% oxygen and 1% carbon dioxide and noble gases.
- Too much atmospheric CO_2 can cause global warming.

Chemical reactions

- **Decomposition reactions** are when compounds are broken down by heat or electricity.
- **Combustion reactions** occur when a substance reacts with oxygen, and they often produce heat.
- **Oxidation** has two definitions:
  A reaction where a substance combines with oxygen.
  A reaction where electrons are removed.
- **Reduction** has two definitions:
  A reaction where oxygen is removed from a substance.
  A reaction where electrons are added.

Rocks and the rock cycle

- **Igneous rocks** are made from magma. The slower the rock is cooled, the larger the crystals found in it.
- **Sedimentary rocks** are formed from rock fragments being deposited and compressed together. Fossils are found in sedimentary rocks.
- **Metamorphic rocks** are formed when rocks are changed by heat and pressure.

The three rock types are linked in the rock cycle:

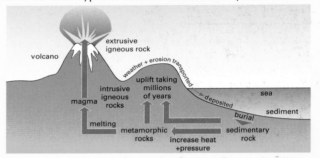

Writing equations

- There are two types of equations: **word and symbol equations**.
- To write a formula you need to know the **combining power of the elements** involved.
- When you have written the symbol equation you must **balance the equation**.

Chemical calculations

- **Relative atomic mass** is the mass of elements relative to carbon 14.
- To calculate the **relative formula mass** of a compound, add up the individual atomic masses of the elements. E.g. the relative formula mass of sodium hydroxide NaOH:
$$23 + 16 + 1 = 40$$
- To calculate the percentage of any element in a compound, use the formula:
$$\% \text{ of element} = \frac{\text{mass of the element}}{\text{mass of compound}} \times 100$$

Forces and motion

- The **speed** of a moving object tells us the rate at which it moves.
- **Acceleration** tells us how the velocity changes per second.
- If the forces on an object are **balanced** the object moves at a **constant speed** in a straight line or remains at **rest**.
- An **unbalanced force** on an object causes it to accelerate:
  - the bigger the force, the bigger the acceleration
  - the bigger the mass, the smaller the acceleration.
- The distance a car travels while the brakes are applied depends on the **braking force, mass** of the car and its occupants, and its **speed**.
- The distance a car travels before it stops depends on the **speed** of the car, the **driver's reactions**, the **braking distance** and the **condition** of the car and the road.
- When an object falls through air, **air resistance** acts on it. The amount of air resistance depends on **how fast** the object is falling and its **shape**. The bigger the weight of the object, the faster it has to be falling to reach **terminal velocity**.

weight	=	mass	×	gravitational field strength
(N)		(kg)		(N/kg)
W	=	mg		

Electricity and magnetism

- **Electric current** is a flow of charge, measured in **amps (A)**. **Voltage** tells us the difference in the energy carried by the charge between the two points. **Resistance** tells us how difficult it is for a current to pass round the circuit.
- When a current-carrying wire is at **right angles** to a magnetic field there is a force on the wire.
- Moving a magnet into a coil produces a voltage across the ends of the coil. This is **electromagnetic induction**.
- **Transformers** alternating current to create a changing magnetic field to induce a voltage in the secondary coil.

voltage (volts)	=	current (amps)	×	resistance (ohms)
electrical power (watts)	=	voltage (volts)	×	current (amps)
energy transferred (joules)	=	power (watts)	×	time (seconds)
energy transferred by an appliance (kilowatt-hours, kWh)	=	power (kilowatts, kW)	×	time (hours)

Waves

- Waves transfer energy without transferring matter.
  Transverse waves: the oscillation is at right angles to the direction in which the energy travels.
  Longitudinal waves: the oscillation is in the same direction as the direction the energy is carried.
- Electromagnetic waves are transverse waves and can travel through a vacuum. The **electromagnetic spectrum** is a set of waves that travel at the same speed in a vacuum.
- **Diffraction** happens when waves pass through a narrower gap or past the edge of a solid barrier.

- Sound is carried by the particles in the medium vibrating, so sound cannot travel through empty space. The **pitch** of a musical note depends on the **frequency** of the vibration.
- The **Earth has a layered structure**. The thin outer layer is broken into large plates that move over the surface. Where the plates move together mountains may fold, volcanoes erupt and earthquakes occur.

wave speed (m/s)	=	frequency (Hz)	×	wavelength (m)
v	=	$f\lambda$		

Energy

- Energy is transferred when there is a temperature difference between two bodies, i.e. when one is hotter than the other.
- Much of the energy we use comes from non-renewable resources, including fossil fuels, e.g. coal and natural gas.
- Renewable sources of energy are those that are continually replaced, like the Sun, wind and waves.

power (watts) (watts)	=	rate of transfer of energy (joules / second)	
work done (joules)	=	force (newtons)	× distance (metres)
PE	=	mgh	
KE	=	$\frac{1}{2} mv^2$	

Radioactivity

- **Radioactivity** is a random process that takes place in the nuclei of some elements.
- **Alpha (α) particles** are made of two protons and two neutrons. **Beta (β) particles** are electrons. **Gamma (γ) radiation** is short wavelength electromagnetic radiation.

- Radiation **ionises molecules** in the material it passes through. Radiation can damage cells in the body.
- Background radiation comes from natural sources and man-made sources in the environment.
- The **half-life** of a radio-isotope is the time taken for half the nuclei present to decay.

Earth and beyond

- The solar system consists of the **Sun, nine planets**, the **asteroid belt** and a number of **comets**. All the bodies in the solar system are held in orbit by gravity.
- The Sun is a star in the Milky Way galaxy, which is one of many millions of galaxies in the Universe.

- Stars are formed when the force of gravity pulls clouds of dust and gas together.
- All the galaxies in the Universe are moving apart very quickly. The distant galaxies, with the bigger red shift, are moving away faster than those nearer to us. This suggests that the Universe was formed by a big bang, which threw all the matter out in different directions.